00-HHE-5C

IDENTIFYING AND CONTROLLING ODOR IN THE MUNICIPAL WASTEWATER ENVIRONMENT

HEALTH EFFECTS OF BIOSOLIDS ODORS: A LITERATURE REVIEW AND ANALYSIS

by:

William S. Cain, Ph.D.
School of Medicine, University of California, San Diego

J. Enrique Cometto-Muñiz, Ph.D.
School of Medicine, University of California, San Diego

2004

WERF

The Water Environment Research Foundation, a not-for-profit organization, funds and manages water quality research for its subscribers through a diverse public-private partnership between municipal utilities, corporations, academia, industry, and the federal government. WERF subscribers include municipal and regional water and wastewater utilities, industrial corporations, environmental engineering firms, and others that share a commitment to cost-effective water quality solutions. WERF is dedicated to advancing science and technology addressing water quality issues as they impact water resources, the atmosphere, the lands, and quality of life.

For more information, contact:
Water Environment Research Foundation
635 Slaters Lane, Suite 300
Alexandria, VA 22314-1177
Tel: (703) 684-2470
Fax: (703) 299-0742
www.werf.org
werf@werf.org

This report was co-published by the following organization. For non-subscriber sales information, contact:

IWA Publishing
Alliance House, 12 Caxton Street
London SW1H 0QS, United Kingdom
Tel: +44 (0) 20 7654 5500
Fax: +44 (0) 20 7654 5555
www.iwapublishing.com
publications@iwap.co.uk

Library of Congress Catalog Card Number: 2004113386

IWAP ISBN: 1-84339-708-0

Organizations that helped prepare this report: UCSD, LACSD, CH2M HILL

ACKNOWLEDGMENTS

The co-principal investigators for WERF Project 00-HHE-5C wish to acknowledge and thank Drs. William S. Cain and J. Enrique Cometto-Muñiz of the School of Medicine at the University of California, San Diego (UCSD) for their in-depth study, analysis, and writing of this addendum report at the UCSD Chemosensory Perception Laboratory in La Jolla, California.

Report Preparation

Co-Principal Investigators
Gregory M. Adams
Los Angeles County Sanitation Districts

Jay Witherspoon
CH2M HILL

Project Team:
William S. Cain, Ph.D.
J. Enrique Cometto-Muñiz, Ph.D.
School of Medicine, University of California, San Diego

J. Ronald Hargreaves
Terence Larro
Los Angeles County Sanitation Districts

Project Subcommittee

Michael Jawson, Ph.D.
U.S. Department of Agriculture – Agricultural Research Service (Chair)

Andrew Chang, Ph.D.
University of California, Riverside

Jane Forste
Jane Forste Associates

Jerry Hatfield, Ph.D.
U.S. Department of Agriculture – Agricultural Research Service National Soils Tilth Lab

Lynn Szabo
DuPont Engineering

John Walker, Ph.D.
U.S. Environmental Protection Agency

Water Environment Research Foundation Staff

Director of Research: Daniel M. Woltering, Ph.D.
Project Manager: Lola Olabode, M.P.H.

ABSTRACT AND BENEFITS

Abstract:

This report deals with whether the experience of odors (i.e., odors as sensations) from biosolids at wastewater treatment plants (WWTPs) causes illness. There exists no repository of information on the numbers of complainants with illness, their specific complaints, or the relationship between degree of exposure and complaints. Nevertheless, anecdotal reports imply a pattern much like that associated with other industrial malodors. Any connection between odor and illness has received little attention in the millions of articles in the medical literature. This state of affairs presumably exists because odors per se generate no objective signs of illness in otherwise healthy persons. Malodors, however, may exacerbate both symptoms and signs of illness in persons with certain chronic disorders, such as asthma and migraine. Vulnerability to such effects varies considerably from person to person.

Symptoms claimed in connection to odors from biosolids in particular seem to come from olfactory rather than chemesthetic (irritating) stimulation, a conclusion derived from calculations that concentrations of emissions of the notable contaminants from WWTPs fail to reach irritating levels even within the grounds of facilities. Any convincing deviation from this expectation would warrant serious attention. Instead, as is true for exposure to malodors from any source, the symptoms associated with WWTP malodors seem to occur via intermediate variables, such as annoyance, anxiety, and frustration. Persons who experience no such distress experience no symptoms. This may prove true for persons with existing illness, but in some cases distress may affect the illness via hormonal mechanisms.

Acknowledgment that odors cause anxiety and the like should inform strategies for dealing with reports of symptoms. Research into the connection between the composition of the emissions from WWTPs and odor characteristics should seek to illuminate quantitative goals that engineers can seek to achieve. Finally, failure to respect the boundary between the subjective and the objective in discussions of the matter can invite flatly incorrect conclusions about the relationship between odors and illness.

Benefits:

- Refines the issues and terminology regarding whether biosolids odors, as sensations, might cause illness;
- Reviews the evidence that biosolids odors do or do not cause illness;
- Explains the functional origins of the acceptability of odors;
- Defines the difference between the sense of smell and the chemesthetic or irritation sense;
- Evaluates whether emissions from WWTPs cause irritation;
- Explains why symptoms attributed to biosolids odors fail to qualify as illness; and
- Clarifies that the amelioration of such symptoms lies in reduction of anxiety and associated states.

Keywords: Biosolids, odors, irritation, illness, symptoms, WWTPs

TABLE OF CONTENTS

LIST OF TABLES

LIST OF FIGURES

ACRONYMS AND ABBREVIATIONS

DMDS	Dimethyl disulfide
DMS	Dimethyl sulfide
MEK	Methyl ethyl ketone
µg S/L	Micrograms (10^{-6} grams) sulfur per liter
ppb	Parts per billion (by weight for liquids, by volume for gases)
ppm	Parts per million (by weight for liquids, by volume for gases)
PubMed	Database of National Library of Medicine
QSAR	Quantitative structure – activity relationship
ug S/L	Micrograms of sulfur per liter
VOC	Volatile organic compound
WWTP	Wastewater treatment plant

Executive Summary

This report addresses the issue of whether odors from biosolids cause illness. Previous discussions have suffered in part from erratic use of terminology on three fronts:

- Lack of clarity regarding symptoms (subjective phenomena) versus signs (objective phenomena) in the definition of illness;
- The distinction between odors as sensations (subjective phenomena) versus the actual odorants and odoriferous emissions (objective phenomena) as the cause of illness; and
- The distinction between the sense of smell (olfaction) versus the chemesthetic sense (sensory irritation).

Prepared to respect the boundaries between the objective and the subjective, and the olfactory and the chemesthetic, the project team surveyed the pertinent literature on emissions from biosolids, giving special attention to quantitative estimations of concentrations of notable constituents and the connection between odors as sensations and illness. Wastewater treatment plants (WWTPs) served as the focus. They receive scrutiny from both the public and regulators. Such scrutiny increases the likelihood of adherence to limits based upon health. In this respect, those limits preclude confounding emissions as threats to health with emissions as the cause of odors. Beyond the survey of the literature, the methodology entailed application of a quantitative structure-activity model to calculate chemesthetic potency and an atmospheric dispersion model to calculate ambient concentration away from a source.

The survey, which included the bibliography of a recent review by Schiffman and collaborators of possible connections between odors, odoriferous emissions, and illness, as well as the 12-million item database of the National Library of Medicine (PubMed), revealed a general lack of attention to whether odors per se cause illness but appropriate attention to the objective toxicological effects of notable contaminants.

The literature's inattention to odors as causes of illness would seem to reflect the absence of signs (objective) to accompany such symptoms (subjective) as nausea and headaches. Nothing in the literature contradicts the appropriateness of the symptoms as part of a continuum of reactions to malodors.

The ability of odor quality to carry emotion-inducing information regarding the condition of the source of malodor reflects the adaptive nature of olfaction. For example, the malodor of spoiled food induces symptoms (e.g., revulsion, nausea) that would likely deter consumption of the tainted food, thereby protecting a person against food poisoning. Nevertheless, the ability of odors to warn against ingestion of bacteria-laden sources has limitations. In particular, olfaction can confuse the danger of contact or ingestion with the danger of inhalation. Perceived odor quality seems to bear no systematic relationship to healthfulness with respect to inhalation. The volatile compounds emitted from biosolids exist in blends of some acceptable, even highly desirable, foods and beverages. That is, the compounds are not inherently bad in either a toxicological or olfactory sense. The relative degree of acceptability depends upon ingredients, upon their proportions, upon context, and upon the meanings people attach to the sensations.

Whereas the terms *pungent* and *acrid* may mean generally unacceptable to the public, they mean irritating (chemesthetic) in the realm of sensory science. Resolution of whether emissions from WWTPs actually cause chemesthetic sensations can come from a comparison of the maximum concentrations measured at WWTPs and estimates of chemesthetic thresholds. Such a comparison indicated that WWTPs do not evoke sensory irritation. Hence, use of terms that imply irritation probably reflect an intent to convey general unacceptability. If credible evidence develops that WWTPs cause irritation, the matter should receive serious attention.

No one can deny that WWTPs may be sources of unacceptable odors even when their emissions fall below levels set to protect public health. The literature gives no encouragement that these unacceptable odors cause illness in healthy persons. In particular:

1. Odors as sensation do not cause signs of illness.
2. The acceptability-unacceptability of odors varies systematically and predictably with circumstances of exposure and depends upon the *meaning* associated with the exposure.
3. Below toxic levels of exposure, symptoms associated with odors involve no pathology.
4. Removal of the source of odor leads to immediate reduction of symptoms.
5. Mediating variables, such as anxiety, seem largely to account for symptoms from odors. The analysis does leave room, however, for a role for malodors in the exacerbation of both symptoms and signs in persons with certain chronic illness, such as asthma. Even if malodors cause effects via an intermediate variable, hormonal processes could then mediate potentiation of symptoms and signs in some patients.

Wastewater treatment plants should take the fifth point to heart — both for the responsibility it implies and for the opportunity it may present for understanding the expression of symptoms. Nothing will reduce symptoms better than control of emissions. WWTPs should, of course, seek to produce as little annoyance or anxiety as possible but should also recognize the mediating states as the source of symptoms.

Research on the sensory consequences of WWTPs could profit from approaches taken in the sensory analysis of foods, for which odor quality receives as much attention as odor intensity and potency. Study of the "sensory anatomy" of odors from WWTPs might sharpen targets for control and might thereby give the engineer greater leverage to minimize annoyance. The information would serve a useful purpose only if it could indeed ensure better control of odors at their source. The effects of malodors on the exacerbation of symptoms or signs in persons with existing illness should receive attention in the research laboratory where issues of dose and response and individual differences could elucidate the risk factors for claims of malodor-induced illness.

CHAPTER 1.0

INTRODUCTION

1.1 Complaints of Illness: Symptom vs. Sign

Some people who experience odors from biosolids in the vicinity of wastewater treatment plants (WWTPs) voice complaints about health. These individuals usually voice such complaints episodically as summaries of experience. For instance, participants at public meetings may claim that odors from biosolids have made them ill in the preceding weeks or months. An inspection of logs of complaints from WWTPs in various parts of the U.S. indicated that day to day people express annoyance rather than illness. From the standpoint of the citizen, annoyance and some forms of illness, such as nausea and headaches, may seem not to differ greatly. From the standpoint of public health, however, annoyance and illness do differ.

The objective of this work is to consider whether odors from biosolids cause illness and to make recommendations for work necessary to understand any effects. We define illness as impairment of normal physiological function, affecting part or all of an organism. In pursuit of its objective, this report includes information on how to view illness via symptoms and signs, how to define odors and thereby to elucidate the question of whether odors may cause illness, how the sense of smell functions, how sensory irritation (chemesthesis) differs from smell, how acceptability of odors derives from a role in physiological regulation, whether the medical community has recognized connections between odors and illness, how field studies have contributed to understanding, and how to study the problem in the future.

Air samples taken periodically around many WWTPs and analyzed for their constituents become the input for calculations of risk of chronic and acute health effects of exposure to emissions from biosolids. Model-based calculations of risk have become guardians of public health around WWTPs (CAPCOA 1993). Emissions that exceed permissible levels can require corrective action. Even when the emissions consistently fall within permissible levels, some citizens may complain of emission-induced illness.

Can claims of illness have any credibility if models of risk rule it out? They could if the odors of the emissions can cause illness. The illnesses of concern typically involve non-life-threatening acute symptoms, such as nausea, headache, shortness of breath, and irritation. These symptoms, commonly called nonspecific because they can occur from exposure to many different agents, often have no objective manifestations. In the terminology of medicine, a symptom is a private, subjective experience. A sign is an objective, outward manifestation. If a person feels nauseated, he has a symptom. If a person feels burning in the eyes and the eyes show increased redness, he has both a symptom and a sign. Not every symptom will give rise to a sign, although it may do so eventually. A person with chronic nausea, for example, might eventually lose weight from avoiding food. The distinction between symptom and sign plays a role in interpretation of reports of illness.

One might imagine that if citizens exposed to emissions from biosolids suffer from symptoms, then workers at WWTPs would too. Dispersion modeling (Appendix B) shows that concentration of a malodorant declines rapidly with distance from a source, so that workers on site would have by far the greater exposure. Obviously, if workers had symptoms regularly, they would fail to function normally, yet they do. The common wisdom holds, however, that reactions in the workplace come from healthy persons, who are self-selected for some tolerance to exposure to the agents in question (NRC 2002). The community at large contains some unhealthy persons and persons who are not self-selected for tolerance. (Health-based limits on emissions from WWTPs take this into account.) Insofar as some nonspecific symptoms may manifest themselves in signs only weakly, if at all, the symptoms provide the *prima facie* case for illness.

Neither the symptomatic person nor even that person's physician can just intuit whether the symptoms come from the airborne agents per se or from the odors of the agents. Assuming that the exposure of the complaining person lies below levels known to cause illness, a physician who seeks the cause for the symptoms might lean toward the odors as sensations as the cause. Does the medical literature endorse the conclusion that odors as sensations cause illness?

CHAPTER 2.0

LITERATURE ON ODORS AND ILLNESS

2.1 Sparse Data and Sparse Literature

Recently, Schiffman, Walker, Dalton et al. (2000) reviewed whether exposure to malodors can have an "impact on physical health." They treated the topic rather generically, i.e., they did not seek to distinguish effects of one malodor from another. Such generic treatment reflects the state of knowledge. Environmental odors can differ substantially in their origins, chemical complexity, periodicity, character (odor quality), concentration (perceived intensity), and acceptability. Insofar as odors could cause illness, then these various characteristics should reflect themselves in the nature, duration, and severity of symptoms or signs. Stated simply, the nature of the effect should relate to the characteristics of the cause.

There apparently exists no repository of information on odor-induced complaints for which citizens have sought medical treatment. Nor does there exist a published literature of clinical observations or studies regarding conditions ascribed to exposure to malodors. To illustrate, a search for the terms *environmental odors* and *illness* in PubMed (http://www.ncbi.nlm.nih.gov/entrez/query.fcgi), the database of more than 12 million items of the National Library of Medicine, yielded 36 papers published over the last three decades. The majority of the 36 dealt with the kind of odor intolerance voiced by people with multiple-chemical sensitivity. Such individuals react adversely to pleasant odors, such as perfumes, as much as to malodors; their reactions provide little general guidance. Only one paper dealt with an instance where fugitive emissions from use of the rather vile-smelling fumigant propyl mercaptan led to occurrence of illness among workers. This reference alone had little relevance to the documentation of illness from odors of biosolids since these include mercaptan odor. (Other pertinent search terms failed to yield appreciably different outcomes.)

For frame of reference about a search that yields 36 papers, note the following: The term breast cancer yielded 119,280 papers; asthma, 67,540 papers; stomach ulcers, 18,467 papers. Few of the diseases that the layperson generally recognizes would yield fewer than 10,000 papers. Even a rare disease like anthrax yielded 1,300 papers before October 2001. The term *extrasensory perception* yielded 571 papers (in a medical database), 16 times as many as *environmental odors* and *illness*. The virtual absence of a published record on odors and illness makes a statement about how the worldwide health community has viewed any claims: It has viewed them as being of little moment.

For example, a search of *hydrogen sulfide* and *toxicity* yielded 1,050 papers. A search of *mercaptan* and *toxicity* yielded 3,193 papers. It is the virtual absence of a literature on illness and odors that tends to direct discussion to the *possibility* of a connection rather than to its nature, to the generic rather than to the specific. Empirical data pertinent to the generic question come principally from a few field studies where people have had an opportunity to voice complaints about odor pollution in their vicinity (see below).

2.2 Need for Resolution

The dearth of literature on illness and odors provides only a circumstantial case that odors as sensations, whether from the biosolids of WWTPs or from other industrial operations, might cause illness. When people experience a symptom they assign it a level of importance for whether it merits a visit to a physician. Perhaps symptoms from odors fail to reach the requisite level of importance. Hence, the case remains open for consideration of other evidence. This situation, we should note, pertains to the general question of whether emissions from biosolids, odorous or inodorous, cause illness. As the National Research Council Committee on Toxicants and Pathogens in Biosolids Applied to Land (NRC 2002) noted, "There have been several allegations of human deaths and illnesses caused by land application of biosolids. However, there has been no documented scientific evidence to substantiate those claims." (p. 52)

The lack of resolution about illness from the odors of biosolids has consequences for WWTPs. Because these facilities exist for the public good, they cannot defend themselves against the charge that they do harm as long as there remains uncertainty about whether their odors cause illness. Some citizens may continue to allege that odors have caused illness no matter what the outcome of scientific studies, but a good-faith effort to examine the matter should allow honest conclusions about cause and effect.

WWTPs need to comply with the same air pollution regulations as any other facility. If the facilities did cause illness while their emissions remained within the limits imposed by the law, then the limits would need scrutiny. As things stand in most jurisdictions, two sets of limits exist: one based upon health and specified in terms of ambient concentration chemical by chemical, and one based upon annoyance and specified in terms of an odor level irrespective of the composition of the emissions. The level based upon odor virtually always requires the more stringent control. Inevitably, though, some citizens will experience malodors at emission levels considered too low to pose a known risk to health.

CHAPTER 3.0

THE MODALITIES

3.1 The Olfactory Process

Discussion of whether malodors induce illness can benefit from knowledge of how the sense of smell functions. The stimuli for smell, or odorants, comprise in almost all instances volatile organic compounds (VOCs) (Cain 1988). The most notable exceptions in everyday life include the inorganic compounds hydrogen sulfide (rotten egg smell), ammonia (not its irritation, but its odor), and ozone.

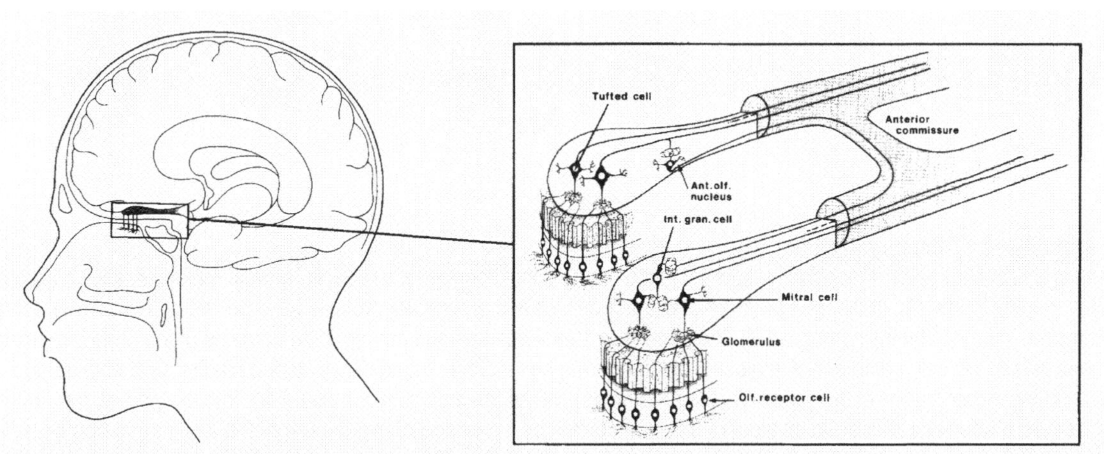

Figure 3-1. Neural Structures of Olfaction.
Olfactory receptor cells reside in a patch at the top of the nasal cavity. Each patch, one in the right cavity and one in the left, contains millions of receptor cells densely packed together. A receptor begins its six- to eight-week life in the mucosal tissue of the cavity and then grows a fiber (axon) through a perforation in bone into the olfactory bulb, a projection of the brain. Each of the perforations shown in the inset would have many fibers coursing through it. The millions of fibers from receptor cells connect to many fewer mitral cells, a few thousand in number. Fibers from the mitral cells pass the neural information along to the next stage of processing in a structure called the pyriform cortex, one of various structures involved in the processing of olfactory information. From there, fibers feed into structures that process emotions and structures involved in regulation of bodily functions, such as eating and drinking. From Ottoson, D. *Physiology of the Nervous System*, p. 435. 1983. Reproduced with permission of Palgrave Macmillan.

The process of smelling begins when molecules of airborne vapor reach olfactory receptor cells (also called receptor neurons) in a patch of tissue called the olfactory mucosa high in each nasal passage (Figures 3-1 and 3-2). The location of the receptor cells enables them to sense odorants in the air during inhalation, as occurs when a person breathes the air around him, and during exhalation, as occurs during eating. Much of the flavor of food comes from olfactory stimulation occasioned by movements of the mouth that send small currents of air up through the nasopharynx, the upper part of the mouth that connects to the nasal cavities.

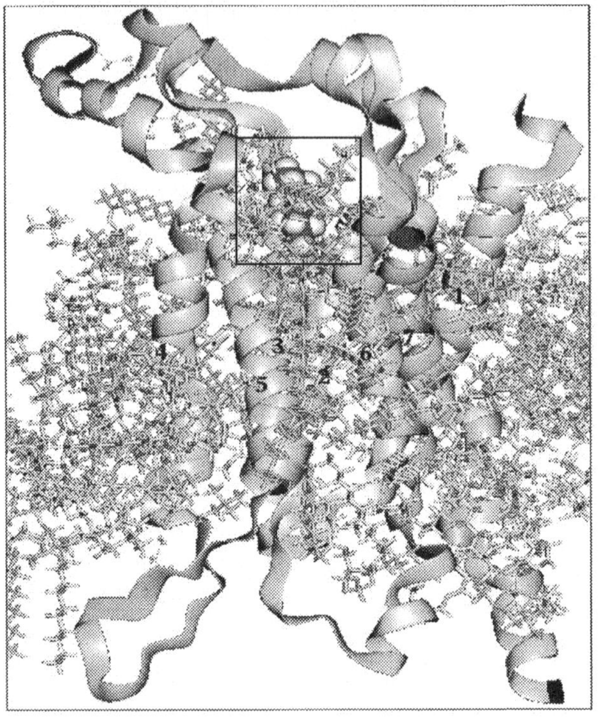

Figure 3-2. An Olfactory Receptor.
Reception occurs when a molecule of odorant lands on the critical portion of a receptor protein molecule, shown here as a ribbon that courses in and out of the membrane of a receptor cell. Olfactory receptor proteins have seven such transmembrane domains, indicated by the numbers 1-7. The fern-like material interspersed amongst the domains of the protein represent the molecules of the cell membrane. A molecule of the odorant hexanol has found its way to the binding pocket at the foot of the arch in the upper portion of the schematic (see rectangle). When the molecule of hexanol fits into the pocket, the protein molecule changes its shape, which triggers a flow of current by a cascade of events. Reprinted with permission from Floriano, W.B., N. Vaidehi, W.A. Goddard III, M.S. Singer, and G.M. Shepherd. 2000. Molecular mechanisms underlying differential odor responses of a mouse olfactory receptor, Proc Nat Acad Sci 97:10712-10716. Copyright (1997) National Academy of Sciences, U.S.A.

Each patch of mucosa contains millions of receptor cells. Each cell contains a certain form of signaling protein embedded in the membrane of its cilia, 12 to 20 hairlike projections at the tip of the cell. A small change in structure that occurs when an airborne molecule of odorant lands on the surface of that protein triggers the process of transduction, the conversion of the chemical energy from the odorant to the electrochemical energy of the nervous system. Approximately 350 different functional signaling proteins occur among the millions of receptor cells, one type per cell. Research has yet to uncover how the protein recognizes the molecule of odorant.

The perceived quality of an odorant depends upon differential binding of its molecules across the array of receptor cells. The differential binding gives rise to a spatial pattern of neural activation. Different odorants create different patterns. The minimum concentrations at which odorants can stimulate depend principally upon the affinity between molecules of odorant and the signaling proteins. This can differ by more than a million-fold. Generally, the materials that smell malodorous seem to have high affinity for their array of signaling proteins. (Various neutral and pleasant smelling vapors also have such high affinity, so high affinity does not define the malodorous domain.) For many malodors, including those from biosolids, perceived quality can become apparent at concentrations below one part per billion (ppb) parts of air and even as low as one part per trillion. Because malodors have such potency as odorants, detection of them often exceeds the sensitivity of instruments.

The odors that people perceive from most sources in the world come from mixtures of chemicals. The aroma of a fruit, for example, comes largely from a variety of esters, alcohols, and organic acids. The aroma of pinewood comes largely from a variety of terpenes and other hydrocarbons. The odor of biosolids comes largely from a variety of organic acids, amines, mercaptans, and sulfides. Unlike fruits and wood, which give off their odors from their intrinsic chemical composition, the odor of biosolids develops principally from bacterial degradation of the organic material. That is, the odor derives from the metabolic products of the bacteria. (Emissions from the original material may continue their presence, but the odors of the bacterial products will dominate because of relative olfactory potency.) This has importance because the odor changes as the type and amount of the bacteria change, as the nutrients available to them change, and as conditions of oxygenation change. The situation makes the emissions from biosolids a moving target from the standpoint of chemical analysis. Hence, both because of limitations on the sensitivity of instruments to detect vanishingly small amounts of airborne odorant and because of the shifting chemical composition of bacterially caused emissions, their instrumental characterization goes well beyond the routine.

Whereas analytical instruments normally separate complex mixtures, such as emissions from biosolids, into constituent components, the sense of smell perceives the mixtures as blends. Only when analytical instruments can perform integration like that which occurs in the nose, can they truly simulate the sense of olfaction. A limiting aspect of tackling malodors from WWTPs is absence of an instrument that could register a quantity that correlates one-to-one with the human olfactory response, even if enough material existed in the air. If such an instrument materialized, it could give the engineer who seeks to control emissions an odor-relevant target to meet (see Defoer, Bo, Langenhove et al. 2002). Engineers solve problems well when they have such a target.

Why do odors vary in acceptability? This seems to derive from a role in the evaluation of what people might eat. Both the sense of smell and its companion modality taste have circuits in the brain to process their messages with respect to whether they meet the needs of internal environment, such as the maintenance of body mass and fluid. These circuits modulate input so as to make it motivating, e.g., to find the aroma of food more inviting during hunger. Acting via that regulatory mode, the modalities both nurture and protect. To wit, people commonly find especially aversive those odors that in general signal the activity of bacteria. Sources for such activity include rotting or decaying meat, fish, rotting vegetable products, feces, and souring dairy products. Although the material that gives rise to the odor may be unhealthful, the associated VOC or odorant commonly is not, at least at levels of environmental relevance. The odorant is just the messenger.

The unacceptability of malodors does not inhere in the odorants. These also exist in nonmalodorous blends. The chemicals in malodors may appear in chocolate, coffee, cheese, cooked meats and fish, beer, wine, fruits and vegetables, and many other products. In a given product, the chemicals exist in small amounts, along with scores or hundreds of others, but they contribute to its essential aroma or flavor. Food chemists in search of active flavor ingredients have had many such surprises when they have separated extracts from natural products into their constituents. Obviously, both ingredients and blend hold the key to the acceptability of an aroma. The sense of smell sometimes can and sometimes cannot discern the constituents in a blend, but can often register their absence. Fish would not smell or taste like it does without the fishy-smelling amines. Most cheese would not taste like it does without putrid-smelling carboxylic acids. Beer would not taste like it does without skunky-smelling mercaptans. In their appropriate blends, they enhance appeal. Hence, there is nothing intrinsically wrong with the fishy-smelling amines, putrid-smelling carboxylic acids, skunky-smelling mercaptans, or most other materials in emissions from biosolids. When these predominate in a blend, however, they send a message of the unacceptability of the source.

Just as acceptability varies depending upon whether a note of rancidity comes from a cheese, from a dirty sock, or from an emission from a plant, acceptability often varies markedly with situation and context. Members of one culture define as delicious flavors that members of other cultures might find disgusting. The adage, "One man's meat is another man's poison" acknowledges both the strength of preferences and the lability of chemosensory experience. Virtually everyone has revised his or her acceptability of chemosensory experiences, more commonly in the positive than in the negative direction. Most adults enjoy some foods they disliked as children. So, not only do people differ from each other, as the adage implies, but they differ from themselves from one time to another. The unknown author of the centuries-old adage saw the irony in how strongly people hold their preferences for, as everyone knows, no man's meat is, in fact, another man's poison. It may just seem that way.

Examples of the situational determinants of acceptability abound. A person who swims in an indoor pool may find the chlorine odor tolerable there but choking in his neighborhood. A person might find the odor of a local stable nauseating until he takes up riding. Does changed acceptability for one odor generalize to all malodors? Seemingly not (Cain & Johnson 1978); it seems to occur case by case.

Dramatic examples of situational determinants of acceptability come from the workplace. In Studs Terkel's (1972) well-known book *Working*, he reported an interview with a mechanic at a rendering and glue factory, who said (p. 111): "The odor was terrible, but I got used to it. It was less annoying when you stayed right in it. When you left for a week or so, a vacation, you had to come back and get used to the thing all over again." These words summarize the experience of many workers in many facilities. Workers in offensive industries do not claim that odors per se have caused illness.

Scientists have not studied why employees in a malodorous industry find the odors to become less objectionable over time while members of the public exposed to the same material, though at much lower intensity, may find the odors to become more objectionable. Part of the reason must concern *meaning*. The odors *mean* something different to the worker than to the member of the public. The worker has voluntary exposure and the member of the public involuntary exposure. The worker gets paid during exposure, a tangible benefit. The member of the public does not. For the worker, the meaning of the exposure is clear. For the member of the public, the meaning can be confusing and provoke anxiety and stress.

3.2 The Chemesthetic Process

It has become common within the last two decades for Americans to experience salsa and related products that add frank piquancy to food. In this respect, Americans and Western

Europeans, too, have caught up with the majority of people in the world, who have long sought frank piquancy through curry powder, paprika, chili flakes, whole hot peppers, Chinese mustard, and the like. Such products, along with the traditional mustards, horseradish, and Tabasco sauce, used to a degree in Western culture, exert their effects through a sensory system known as chemesthesis, or the capacity to feel the presence of chemicals. Because the sensations of chemesthesis occur so commonly in the area of the face, some scientists talk about chemesthesis as the trigeminal system, after the nerve that mediates facial chemesthesis (Figure 3-3). In fact, though, the trigeminal nerve also mediates non-chemosensory information, such as cold, warmth, touch, and mechanically induced pain. Furthermore, other nerves in the tracheal, genital, and anal areas also mediate the feel of chemicals. So, using the term *chemesthesis* is more appropriate than referring to the trigeminal system.

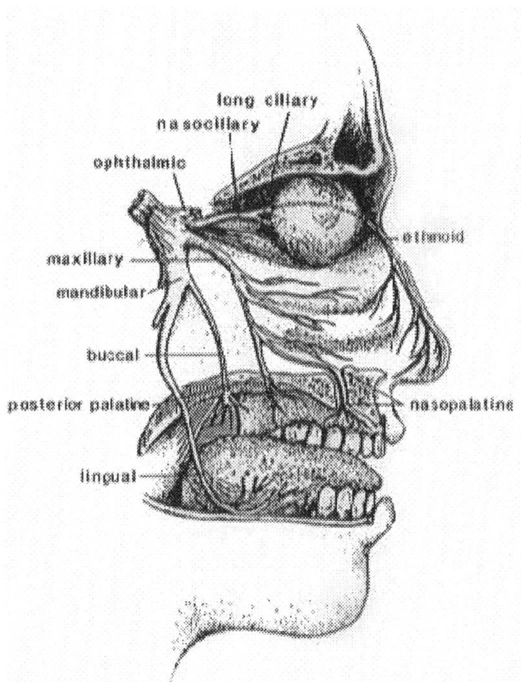

Figure 3-3. Anatomy of Trigeminal Innervation.
The various branches of the trigeminal nerve mediate all somesthetic sensation in the face (touch, warmth, cold, pain), including chemesthetic sensations (pungency). In the posterior oral cavity and laryngeal areas, two other nerves, glossopharyngeal and vagus, serve the role of the trigeminal nerve. Reprinted by permission of Finger, T.E., W.D. Silver, and D. Restrepo, 2000. Neurobiology of Taste and Smell, New York, Wiley-Liss, Inc., a subsidiary of John Wiley & Sons, Inc.

Some vapors have notable properties that evoke nasal chemesthesis. These include vapors of ammonia, oven cleaner, and many other cleaners. Sometimes, people take the chemesthesis as a sign of the efficacy of a product to clean. In fact, most vapors (e.g., solvents, such as turpentine) have the capacity to stimulate chemesthesis at high concentrations. If the concentration of such a material rises from very low to high, the material will first reveal its characteristic odor. As the odor grows progressively stronger, the vapor will begin to cause a feel or irritation in the nose and eyes (Cometto-Muñiz, Cain, Abraham et al. 1998). Just because a chemical can stimulate chemesthesis does not mean that it will. The issue becomes quantitative. Vapors vary tremendously in their potency depending on the properties of their molecules (Cometto-Muñiz and Cain 1994, 1996; Cometto-Muñiz 2001). Fortunately, certain models can predict the potency of vapors to stimulate chemesthesis from properties of their molecules (Abraham, Kumarsingh, Cometto-Muñiz et al. 1998a,b; Abraham, Gola, Cometto-Muñiz et al. 2001; Abraham, Hassanisadi, Jalali-Heravi et al. 2003).

CHAPTER 4.0

POTENTIAL EFFECTS VS. ACTUAL EFFECTS

4.1 Focus on the Sensory

Despite the general absence of a literature that demonstrates that odors cause illness, one may still ask, "Do odors have the potential to cause illness?" Schiffman et al. (2000) addressed that question with respect to odors from animal operations, wastewater treatment, and recycling of byproducts in a workshop convened in 1997. The scientists seemed to answer the question *yes*. As posed, the question essentially allows little room for a categorical answer of *no*. No scientist would reject the possibility that some odor might in some circumstance cause a health effect, even if the probability seemed small. Beyond that consideration, the answer requires a somewhat sophisticated interpretation, beginning with the ground covered by the deliberations of the scientists. In their report, the authors mentioned "three paradigms by which ambient odors may produce health symptoms in communities with odorous manures and biosolids. In the first paradigm, the symptoms are induced by exposure to odorants at levels that also cause irritation (or other toxicological effects)....In the second paradigm health symptoms occur at concentrations that are not irritating....In the third paradigm, the odorant is part of a mixture that contains a co-pollutant that is essentially responsible for the reported health symptom" (p. 9–10). (The reader can safely substitute the word *example* for *paradigm,* as used by Schiffman et al.)

Regarding the first paradigm, if the symptoms come from sensory irritation (chemesthesis), then this aspect of the vapor, rather than its odor, bears responsibility for the effect. In practical terms, this means that the effect must almost certainly occur at high levels of stimulation, generally a hundred times or more above that for odor (Cometto-Muñiz & Cain 1995). Similarly, if another toxicological effect induces the symptoms, then that effect, rather than the odor, bears the responsibility. Knowledge of the system affected holds the key to solution of the problem.

Regarding the second paradigm, this example pertains more closely to the issue of odors as sensations, the topic of concern here. If odors, properly considered as internal representations of odorants, cause health effects, then they do so as sensations, much as the sight of a car wreck may make a person feel ill. The wreck does not make the person feel ill, the sight of it does.

Regarding the third paradigm, if symptoms occur because of a co-pollutant rather than odor, then as for the first paradigm the toxic property of the co-pollutant should bear the responsibility. Blaming the odor diverts attention from the true problem.

How should we turn such considerations into action? First, we should dismiss paradigm three as irrelevant here because it represents cases of mistaken identity. If malodorous pollution causes illness because of other travelers, the likely culprit is viable particulate matter, a common form of pollution from animal facilities, but a negligible form from WWTPs (Bottcher 2001).

Second, we should dismiss the parenthetical "other toxicological effects" from paradigm one, but we can retain irritation. Although irritation occurs by means of a modality other than olfaction, i.e., by means of chemesthesis, it would seem evasive not to agree that from the standpoint of the citizen, the odor from a material and the irritation from that same material admit to no simple phenomenological boundary. Fortunately, the information available regarding irritation from VOCs allows us to address this issue quantitatively. Regarding "other toxicological effects," we argue to dismiss it as separate from odor or irritation, for that matter.

Third, we should accept paradigm two as central to our concerns. If malodors, as odors, from biosolids cause illness, then we should deal with the emissions as injurious to public health. If this outcome fails to hold, then we should deal with the malodors as sources of annoyance. Criteria based upon public health and criteria based upon annoyance may, as indicated above, take different forms. A criterion based upon public health would most likely take the form of a fixed physical limit on exposure. A criterion based upon annoyance has historically taken a more flexible form in which expressions of annoyance trigger enforcement. Hence, if no one complains, no action occurs. This historical situation has, however, begun to change, particularly in Europe with introduction of more rigid regulations on odoriferous emissions (Mahin 2001).

To summarize, when one avoids the confounding indicated above, there remain just two central questions:

- Do the emissions from WWTPs rise to a level that triggers irritation?
- Do the odor sensations from the emissions cause health effects? The other questions embodied in the three paradigms do not address matters of the sense of smell.

4.2 Generality vs. Particularity

To approximate the answer to the question of whether odors from biosolids cause illness, we can ask whether industrial malodors *in general* cause health effects. (This question differs from the question posed by Schiffman et al. of whether odors *could* cause health effects. Moreover, this question could lead to the simple answer *yes* or *no*, whereas their question may forever preclude the answer *no*.) Malodors caused by various industrial sources often have constituents in common, e.g., the rotten-egg smell of hydrogen sulfide and the putrid smell of carboxylic acids. Pulp and paper factories put out many of the same sulfides as WWTPs. Rendering plants put out many of the same carboxylic acids as WWTPs. If the evidence suggests that odors from these various sources cause health effects, then one should refine the question to focus on odors from biosolids and deal with the toxicological particulars: What specific health effects? Which odors from biosolids? At what level of stimulation? Over what duration of exposure? Surely, if odors cause illness, they would do so by certain principles.

4.3 Chemesthetic Effects

As noted above, a person will undoubtedly see any distinction between odor and irritation as meaningless when either can come from the same material. When people describe malodors, they may use such terms as pungent or acrid, perhaps inappropriately to mean foul smelling or annoying, rather than actually irritating. That use, appropriate or inappropriate, provides incentive to examine whether measured concentrations of emissions from biosolids can indeed stimulate chemesthesis. If citizens experienced actual chemesthetic sensations on a regular basis, they might well develop verifiable signs of the exposures as well as symptoms. The signs would likely include reddened eyes and rhinitis (nasal congestion and runny nose) that should outlast the stimulation.

Chapter 4.0 of the Final Phase 1 Report for WERF Project 00-HHE-5, *Identifying and Controlling Odor in the Municipal Wastewater Environment*, established a list of the top 10 odorous compounds or groups of compounds (shown here alphabetically) consistently found to emanate from biosolids at WWTPs:

1. Amines, particularly trimethyl amine, with a fishy-urinous odor quality;
2. Ammonia;
3. Carbon disulfide, and other organo-sulfides, with flatulent odor quality;
4. Carboxylic acids, particularly, propionic and butyric acids, with putrid-rancid odor quality;
5. Dimethyl disulfide, with a skunky, burnt-feather odor quality;
6. Dimethyl sulfide, with a radish-cabbage odor quality;
7. Hydrogen sulfide, with a rotten egg quality;
8. Ketones, particularly methyl ethyl ketone and acetone, with an ethereal odor quality;
9. Mercaptans, particularly methyl mercaptan, with a skunky, dead-animal odor quality; and
10. Terpenes, particularly, C_{10} compounds such as limonene, and alpha- and beta-pinene, with citrus, woody qualities.

The ketones and terpenes do not have unpleasant odor qualities, so these obviously play little role in the net odor quality of the effluents. The ketones may occur in odors from WWTPs because they are solvents. Acetone, for example, is used as nail polish remover. The terpenes serve as solvents, e.g., turpentine, but also occur as fragrances, e.g., citrus-smelling limonene adds fragrance to household cleaners, and they occur in fresh cuttings of shrubbery. Clearly, the unpleasant-smelling sulfur-containing compounds, the amines, and the carboxylic acids account principally for the negative odor character.

We sought, for the various compounds, information regarding concentrations in and around WWTPs. Turning to the article by Schiffman et al. (2000), to databases, such as PubMed, and to reference lists of articles retrieved, we found more than 100 articles of interest. Only 10, however, contained quantitative information regarding concentrations. Table 4-1 contains that information in terms of the following: (1) the industry that releases a contaminant; (2) the environment or process that generated the material; (3) the concentration measured; and (4) the reference. Only when equipped with information on concentrations could we ask whether these materials might evoke chemesthesis.

As mentioned, concentrations that evoke chemesthesis usually lie orders of magnitude above concentrations that evoke odor (see also Ruth 1986; Devos, Patte, Rouault et al. 1990). Accordingly, chemesthesis would generally occur only in the presence of overpowering malodor.

Table 4-1. Concentrations (ppb) of Odorants at Various Sources.				
Odorant	**Industry**	**Environment/Process**	**ppb**	**Reference**
Acetone	Sewage		0.92	Rosenfeld et al. 2001
	Sewage treatment plants (12)	Thermal conditioning	790	Bonnin et al. 1990
Ammonia	Sewage treatment plants (12)	Sludge storage	28,933	Bonnin et al. 1990
	Sewage plants (16) and pumping stations (18)	Sludge treatment	5,060	Kangas et al. 1986
	Biosolids composting facility	Inlet of packed tower odor scrubbing system (peak)	300,000	Muirhead et al. 1993
	Sludge composting system	Exhaust of odor control unit	7,000	Horst et al. 1991
Butanone (MEK)	Sewage		0.15	Rosenfeld et al. 2001
Carbon disulfide	Sewage		0.054	Rosenfeld et al. 2001
Dimethyl disulfide	Sewage		0.088	Rosenfeld et al. 2001
(DMDS)	Sewage treatment plants (12)	Thermal conditioning	31	Bonnin et al. 1990
	Wastewater treatment plants	Digester dome of floating roof	26.4 μg S/L	Devai et al. 1999
	Sludge composting system	Exhaust of odor control unit	1,000	Horst et al. 1991
Dimethyl sulfide	Sewage		0.027	Rosenfeld et al. 2001
(DMS)	Sewage plants (16) and pumping stations (18)	Sludge treatment	898	Kangas et al. 1986
(DMS + DMDS)	Biosolids composting facility	Inlet of packed tower odor scrubbing system (peak)	15,000	Muirhead et al. 1993
Hydrogen sulfide	Sewage	Purification plant	100,000	Sastrand et al. 2000
	Organic wastes composting	Vegetable, fruit, garden refuse	500	Heida et al. 1995
	Sewage treatment plants (12)	Belt press working	7,336	Bonnin et al. 1990
	Wastewater treatment plants	Digester dome of floating roof	446.9 μg S/L	Devai et al. 1999
	Sewage plants (16) and pumping stations (18)	Screens	1,353	Kangas et al. 1986
Limonene	Organic wastes composting	Vegetable, fruit, garden refuse	24,944	Heida et al. 1995
	Biowaste	Landfill (surface of windrow)	226,453	Tolvanen et al. 1998
	Biowaste	Landfill (inner part of windrow at 2 weeks)	34,939	Tolvanen et al. 1998
Methyl mercaptan	Sewage treatment			
	Sewage treatment plants (12)	Belt press working	1,564	Bonnin et al. 1990
	Wastewater treatment plants	Digester dome of floating roof	8.7 μg S/L	Devai et al. 1999
	Sewage plants (16) and pumping stations (18)	Sludge treatment	26,739	Kangas et al. 1986
(mercaptans)	Biosolids composting facility	Inlet of packed tower odor scrubbing system (peak)	2,000	Muirhead et al. 1993
	Sewage		121	Rosenfeld et al. 2001
Pinene (alpha)	Biowaste	Landfill (surface of windrow at 1 week)	2,084	Tolvanen et al. 1998
	Biowaste	Landfill (inner part of windrow at 2 weeks)	9,638	Tolvanen et al. 1998
Pinene (beta)	Biowaste	Landfill (surface of windrow at 1 week)	1,183	Tolvanen et al. 1998
	Biowaste	Landfill (inner part of windrow at 2 weeks)	2,904	Tolvanen et al. 1998

For frame of reference, a malodor of the sort emitted from biosolids normally becomes quite objectionable for its odor quality at a concentration just a small multiple of its odor threshold, commonly three to seven. Although this consideration made it unlikely that the VOCs reached their chemesthetic thresholds, we made the quantitative comparisons where possible.

A reliable data set on nasal chemesthetic thresholds contains only a few dozen chemicals (see Cometto-Muñiz 2001). This set consists of thresholds uninfluenced by odor, gathered either from persons who have no sense of smell or by a technique called nasal localization that bypasses the distracting influence from the odor of the materials. Despite its relatively small size, the data set has lent itself to erection of a quantitative structure-activity relationship (QSAR) based upon physicochemical properties (Abraham et al. 1998a,b; Abraham et al. 2001; Abraham et al. 2003).

The QSAR becomes important in the present case because the data set contains chemesthetic data for only four of the materials of interest (acetone, butyric acid, limonene, and alpha-pinene). For the others, potency needed to be calculated by the equations in Abraham et al. (2001). These equations were built from VOC data, not from data on inorganic volatiles. Hence, its applicability to compounds such as ammonia and hydrogen sulfide does not rest on firm ground. For these two substances, we have relied on values provided in the literature (Ruth 1986) even though such chemesthetic thresholds are most likely confounded by odor.

Table 4-2 shows measured and calculated chemesthetic thresholds for the materials. For the 13 chemicals in the table, all but two — ammonia and hydrogen sulfide — exceeded the concentrations measured in the environment by large factors. This outcome is all the more striking because the environmental concentrations came principally from measurements in the facilities, and even in the heart of a process, not from the community. No member of the public would ever have exposure to these concentrations. Furthermore, in the case of hydrogen sulfide the value (100 ppm) represents a maximum weekly peak excursion of 14–40 minutes, potentially significant only inside sludge stores in the sewage plant (Søstrand et al. 2000). In the case of ammonia, the value (300 ppm) represents the peak *inlet* concentration to an odor scrubbing system that at the outlet showed non-detectable ammonia, indicating a > 99.9% removal (Muirhead et al. 1993).

Even if the chemesthetic effect of vapors from biosolids arose from the effect of all the constituents in the effluent summed, the ambient concentrations would still fall below the chemesthetic threshold (Cometto-Muñiz, Cain, and Hudnell 1997; Cometto-Muñiz, Cain, Abraham et al. 1999, 2001). Chemesthesis accordingly fails to explain symptoms associated with emissions from WWTPs.

Table 4-2. Thresholds for Irritation, Maximum Airborne Concentrations Around Biosolids, and the Ratio of Thresholds to Airborne Concentrations.

Odorant	Nasal Pungency (ppm)		Eye Irritation (ppm)		Maximum Concentration (ppm)		Ratio Threshold to Maximum Concentration
Acetone	130,671	#	186,265	#	0.79	1	165,406
Ammonia	102	a			300	2	0.34
Butyric acid	61	#	58	*			
Butanone (MEK)	21,380	*	26,303	*	0.00015	3	142,533,333
Carbon disulfide	363,078	*	> vapor saturation	*	0.000054	3	6,723,666,667
Dimethyl	9,120	*	18,197	*	1	4	9,120
Dimethyl sulfide	151,356	*	204,174	*	15	2	10,090
Hydrogen sulfide	10	a			100	5	0.1
Limonene	> vapor	#	> vapor	*	226	6	N/A
Methyl mercaptan	426,580	*	575,440	*	27	7	15,799
Pinene (alpha)	> vapor	#	> vapor	*	10	8	N/A
Propionic acid	324	*	155	*			
Trimethyl amine	147,911	*	316,228	*	20	2	7,396

\# measured value

a from Ruth, 1986

* calculated value

1 Thermal conditioning in sewer treatment plant (indoors) (Bonnin et al. 1990)

2 Inlet to packed tower odor scrubbing system (peak condition) (indoors) (Muirhead et al. 1993)

3 (Sewage) (indoors, inside chambers) (Rosenfeld et al. 2001)

4 Exhaust of odor control unit (outdoors) (Horst et al. 1991)

5 Sewage purification plant (indoors) (Søstrand et al. 2000)

6 Landfill, biowaste (surface of windrow at 1 week) (Tolvanen et al. 1998)

7 Sludge treatment (indoors) (Kangas et al. 1986)

8 Landfill, biowaste (inner part of the windrow at 2 weeks) (Tolvanen et al. 1998)

4.4 Field Studies

There exists a small literature on field studies of odor pollution, in which participants gave their reactions to malodors perceived in their neighborhoods. The sources have varied from composting to oil refineries. Questions commonly ranged from those about mere frequency of perception to those about symptoms. By suitable choice of samples of participants, the investigators could study the pattern of responses for such issues as whether old people have more complaints than young people, whether those who worry about their health have more complaints than those who do not, and so on.

Approximately 50 different types of industries, defined by Standard Industrial Classification code, have odoriferous operations (Flesh, Burns, and Turk 1974). Emissions from these can affect tens of millions of people, a number that hardly escapes notice in the public health community. Air pollution control districts, an arm of public health, typically expend more resources investigating odor complaints than any other type. In most cases, field studies came about from particularly vigorous reactions of communities to a local source of odor pollution (Winnecke and Kastka 1977, 1987).

The field study became a tool in research on environmental odors in the 1960s and '70s (e.g., Cederlöf, Friberg, Jonsson et al. 1964; Jonsson, Deane, and Sanders 1975; Deane & Sanders 1977; 1978; Deane, Sanders, and Jonsson 1977). Since then, understanding of the nature of complaints about malodors has grown incrementally with successive surveys. Somewhat surprisingly, but at the same time reassuringly, the essential findings have held across industries and across surveys.

From the outset, it became apparent that women would generally prove more reactive to malodors than would men, young people would prove more reactive than older people, people with existing chronic disorders would prove more reactive than healthy people, and people with greater proclivity to complain would prove more reactive than people with less proclivity to do so (Jonsson 1974).

In the majority of studies, investigators have concluded that somatic, i.e., bodily, symptoms occur via a nonphysical intermediate variable, such as annoyance, stress, anxiety, or worry, rather than via pathophysiology (Winnecke, Neuf, and Steinheider 1996). For instance, the more annoyed a person from an odor, irrespective of whether the person has high exposure or low exposure, whether the person lives near the source or farther from it, the more likely the person is to express somatic symptoms (Steinhelder, 1998/9; Steinheider, Both, and Winnecke 1998a,b). The phenomenon of worry, exemplified in answers to the question, "How worried or concerned are you about environmental hazards in your neighborhood," accounted for symptoms from odors near a closed waste disposal site (Lipscomb, Goldman, Satin et al. 1991; Shusterman, Lipscomb, Neutra et al. 1991). This is reflected over duration of exposure as well. The longer a person experiences a malodor (e.g., months versus days or weeks), the more likely the person is to express somatic complaints. Indeed, symptoms may remain high even when levels of odor have decreased significantly over time (Lipscomb et al. 1991; Luginaah, Taylor, Elliott et al. 2000). This reinforces the conclusion that meaning plays a major role in unacceptability.

The finding that symptoms may occur via an intermediate variable has considerable acceptance in the medical community, as physicians and scientists seek more and more to understand cases of medical symptoms without identified pathology (Katon, Sullivan, and Walker 2001). As Katon et al. note (p. 921) with respect to the large number of visits to physicians where no medical explanation is found: "Distress and disease both produce physical symptoms. It is not productive to dichotomize symptoms as 'somatogenic' and 'psychogenic' because physiologic and psychological processes are involved in all symptom production and perception."

Whether or not that rule has universal relevance, it may well have relevance regarding exacerbation of existing illness. To illustrate, physicians who manage patients with asthma will hear from some patients that stress exacerbates their bronchoconstriction. Studies give some credence to this claim. As Smailing, McKnight, and Afari (2002) found, "Mood and stress may explain small but important proportions of the variance in daily pulmonary function among some people with mild-to-moderate asthma" (p. 509). Hormonal responses, such as level of cortisol, may mediate such effects (Laube, Curbow, Costello et al. 2002).

Physicians also hear that odors exacerbate bronchoconstriction. Historically, this claim pertained to odors from flowers (Eriksson, Lowhagen, Nilsson et al. 1987). Research has not resolved the matter (Shim and Williams, 1986; Millqvist and Lowhagen 1996, 1998). One could nevertheless see how the odors of flowers could induce stress or anxiety in some asthmatics, particularly the many with the allergic trigger of pollen. Perception of a floral odor could seem to mean exposure to pollen, the true allergen, even when it does not. Whereas this situation could not induce bronchoconstriction in normal people, it could via hormonal responses in asthmatics. The same could well prove true with respect to malodors, such as those emitted from biosolids. Indeed, Segala, Poizeau, and Mace (2003) found in a field survey (via telephone) of a WWTP that people with existing respiratory illness reported more severe symptoms than normal persons.

Might malodors exacerbate other conditions, if only temporarily? Migraine provides another ready example where odor may lead to stress that may in turn exacerbate the existing illness (Wacogne, Lacoste, Guillibert et al., 2003; Grosser, Oelkers, Hummel et al. 2000).

CHAPTER 5.0

DRAWING CONNECTIONS

5.1 Odors and Illness

In most jurisdictions, the limit for nuisance from emissions lies between two and seven times the odor threshold, very far below levels of biological or toxicological concern (Mahin 2001). Only further investigation could reveal the frequency and nature of claims of symptoms from emissions below levels of concern to health. Any such studies should include the best available measures of exposure, for without knowledge of what people breathed and when, information about their complaints can have little meaning.

What might the studies find? They could find existence of definable illness in otherwise well persons, with signs as well as symptoms in affected individuals. This would indicate a need for tightening existing limits based upon biological effects. Because a field study would lack the precision to understand true cause and effect, the finding would prompt laboratory studies of this previously undetected effect (Shusterman 1999). The likelihood of finding definable illness seems quite remote, though, because limits based upon health criteria already have margins of safety.

Field studies might find the existence in otherwise well persons of a malodor-induced syndrome, manifested as a collection or pattern of symptoms without signs (Shusterman 1999). By the common model of disease, symptoms reflect underlying pathology. The physician learns to relate sets of symptoms to particular pathologies and will normally embark upon testing for pathology. In recent years, physicians have begun to see various syndromes without definable or unifying pathology, such as irritable bowel syndrome, fibromyalgia, and multiple chemical sensitivity. When presented with a pattern of symptoms associated with odor pollution, the physician may find no signs of illness, no dysfunctional system, no organic pathology. Although such a syndrome, should one exist, might therefore share some characteristics with these others, it would differ in the important feature that it disappears when the odors disappear. By customary criteria, conditions that do not outlast the inciting event do not qualify as illness.

Field studies might find no definable illness, no odor-induced syndrome, but just isolated symptoms that may or may not have a causal connection with odors. A person with no prior problem with sleep may find that malodors interfere with the ability to fall asleep.

Most likely, however, the field studies would find a combination of no odor-induced illness in otherwise normal people, but some exacerbation of symptoms, and perhaps signs, in some people with existing illness. People with chronic respiratory illness, with gastric illness, headaches, insomnia, anxiety disorders, and the like, may feel more ill during periods of odor pollution. The findings of previous field studies that an intermediate variable, such as odor-induced anxiety or stress, accounts for the feelings, or more exactly for the exacerbation of the feelings, seem entirely relevant here.

5.2 Focus on WWTPs

This report has focused on emissions from WWTPs for various reasons. Many WWTPs share characteristics, such as operation in a fixed location with many nearby residents over long periods of time during which they need develop a relationship — sometimes cordial and sometimes contentious — with residents. As known sources of odoriferous emissions, WWTPs operate under scrutiny of air pollution control districts, as well of residents, and should keep their emissions within limits both by criteria of annoyance and criteria of health. As part of the scrutiny, there exist measurements of emissions around WWTPs. If people complain that odors around such scrutinized facilities make them ill, it seems highly unlikely that the complaints derive from toxic properties, as we have explained.

We have less certainty about the origin of complaints of illness from biosolids applied to land because the emissions occur with less scrutiny. Hence, a complaint of odor-induced illness could, in principle, occur for reasons other than odor. Aside from that, odor-induced complaints should follow the same rules in both situations.

In this report, we have sought to keep the questions and the terminology straight, for even a little slippage can lend unwarranted encouragement to the conclusion that odors cause illness. To illustrate, in the "Final Comments" of the report by Schiffman et al. (2000), the authors state, "Our current state of knowledge clearly suggests that it is possible for odorous emissions from animal operations, wastewater treatment, and recycling of biosolids to have an impact on physical health." This statement, which might appear to encourage the conclusion that odors do cause illness, should not. The term "odorous emissions" makes the statement ambiguous with respect to odors. Whereas the statement might well pertain to the situation in which a toxic property other than odor causes the symptom (Schiffman et al.'s first paradigm) or a co-pollutant causes the symptom (their third paradigm), it fails to address whether odor per se has an effect on health. Without that clarification, readers can draw the wrong conclusion. We do not imply any intent to mislead on the part of Schiffman et al., but we understand the temptation of readers to take away anything they might wish from it.

Does the present report differ from other general accounts, such as those of Shusterman (1992, 1999)? It does in its exclusive focus on the question of whether odors as sensations cause illness. This issue seems to have remained unsettled in spite of the evidence. As discussed with respect to the paradigms of Schiffman et al., all the other issues concern toxicology, not odors. When odoriferous emissions cause illness because of the toxic or carcinogenic properties of odorants themselves or from other travelers, then the fact that the emissions have odor is immaterial from the standpoint of health.

Surely, if a WWTP egregiously violates standards based upon nuisance, it invites complaints. Field studies imply that somatic complaints increase with duration and magnitude of offending odors, irrespective of the industry. Such complaints will, we aver, come predominantly from persons with certain chronic conditions. We cannot rule out the possibility that somatic complaints could also reflect an increase in frustration in healthy people or even an increase in the number of incidents that could possibly be ascribed to exposure to the odors, irrespective of actual cause and effect. As Katon, Sulllivan, and Walker (2001) note, symptoms can serve as a way to express perceived distress and powerlessness. No evidence suggests that exposure to odors has any cumulative effect on physiological systems but surely under protracted conditions of malodor, symptoms may sometimes provide the only source of leverage and power.

CHAPTER 6.0

CONCLUSIONS AND PROSPECTS

6.1 Conclusions

Reasons to conclude that odor sensations do not cause illness in healthy people include the following:

- Odors do not cause signs of illness in the healthy.
- The acceptability-unacceptability of odors varies systematically and predictably with circumstances of exposure and depends upon the *meaning* people associate with the exposure.
- Below toxic levels of exposure, symptoms associated with odors involve no pathology.
- Removal of the source of odor results in virtually immediate reduction of symptoms.
- Nonphysical variables, such as anxiety and stress, seem to mediate symptoms from odors. This does not mean that the symptoms arise inappropriately. In view of the role of smell in protection against hazardous ingestion, symptoms can be appropriate, even if not manifestations of illness.

For people with certain chronic illnesses, exposure to malodors may exacerbate existing symptoms and possibly existing signs. Nonphysical variables may mediate the effects, but through modulation of existing pathology. A role of mediating variables means that the management of people's symptoms lies in reduction of those states. Studies of how to do so might shed as much light on the nature of the reactions as any study of the reactions themselves.

6.2 Prospects

Although we have not focused upon biosolids applied to land, it does appear that most complaints about them concern odor (NRC 2002). It has undoubtedly occurred to everyone associated with biosolids that elimination of their odor might eliminate objections to their use in agriculture. There do exist other areas of contention, such as whether the level of pathogens in biosolids poses risks to health, but the issue of odors seems to confound almost every debate on the pros and cons of use. This would seem to argue strongly for attention to control of the odor.

We have asserted that engineers can solve problems when they see a well-defined target. To understand the target, scientists need to decipher the relationship between the various perceptual aspects of the malodors and physical constitution. Although some research has addressed this issue (e.g., Gostelow, Parsons, and Stuetz 2001; Lambert, Beaman, and Winter 2000), it has lacked the sensory analytical sophistication manifest in food science, where investigators have teased apart both stimulus and sensation. The solution to the problem will require understanding correspondences between the sensory structure of the malodors from biosolids and the chemicals that cause the malodors. Particular sensory problems that need attention include the following:

1. *Collect data on the distribution of threshold sensitivity to the notable constituents of emissions from biosolids:* There exist databases of the thresholds for the odors of various VOCs, but these databases contain information generally unverified with respect to vapor-phase concentration. They often rely upon results collected almost a century ago. They also lack information about the relationship between concentration and probability of detection both within and across individuals. So, a database may contain values of concentrations that 50% of people can detect but contain no information about how much attenuation would be necessary for detection to fall to 10%, 5%, and so forth. Of even greater concern, the databases have notorious levels of error, as high as orders of magnitude.

   Modern olfactometry now makes collection of reliable and environmentally realistic data on threshold detection almost routine. We say "almost" because the requirement of chemical analysis at levels detectable by the nose still goes beyond the routine. Nevertheless, the collection of data on the distribution of sensitivity to even two dozen of the more potent odoriferous materials from biosolids could have considerable value as targets for control.

2. *Supplement knowledge of thresholds for individual chemicals with those for mixtures:* There exist some results on the relationship between the detection of individual chemicals and the detection of mixtures. For emissions from biosolids, the questions of relevance concern (a) whether chemicals with like functionality, e.g., sulfides, show any different rules of additivity than materials of unlike functionality, e.g., sulfides and amines, and (b) whether the rules of additivity of complex mixtures differ from those of simple mixtures.

3. *Above the threshold, perceived odor intensity increases nonlinearly with concentration, with different functions for different materials:* Simply put, for a given reduction in concentration, the odor intensity of one material can decrease more than that of another. Any effort to collect data on the distribution of threshold sensitivity could also collect data on supra-threshold odor intensity for the same materials and for some mixtures.

4. *Research on the intensive properties of malodors needs to be supplemented with research on qualitative properties, i.e., the character and acceptability of the odors, particularly of blends:* This topic has received very little attention in the small amount of research devoted to the sensory analysis of malodors from biosolids.

Research on the four topics identified above should bring understanding that goes beyond the data. Some of the understanding might lie in discovery of physicochemical correlates of sensory phenomena, and some may lie in discovery of the neural "algebra" of olfactory processing. The problems that face the wastewater treatment industry do not differ substantially from those that face other industries with chemosensory concerns, including the fragrance and flavor industry. Understanding the connections among these seemingly disparate fields can allow discoveries in one venue to leverage those in another.

Research on the sensory properties of the odors of biosolids can set the stage for control. Actual control will give rise to its own needs for research, mostly on the chemistry of biosolids, though this could occur as understanding of the sensory properties unfolds.

Should some research focus upon health effects, as well as some on control? We see merit in studies of the effects of the odors on patients with asthma and perhaps with migraine. Such studies could take place in the controlled conditions of a laboratory and could thereby allow exploration of dose and response. These studies could tease apart issues concerning whether chemical agents themselves, their odors as odors, or the stress associated with stimulation govern effects. The studies could also address whether positive effects occur only in some persons and what might determine the extent of any individual differences.

APPENDIX A

FREQUENTLY ASKED QUESTIONS

Why can the air around a wastewater treatment plant smell unpleasant?

A wastewater treatment plant collects the waste that flows through the sewers. Much waste smells bad even as it enters the sewers. Obviously, a place that collects such material has a challenging task of odor control. The wastewater treatment plant, however, not only collects odoriferous waste, but treats it through a series of processes that themselves can cause odor. The goal of treatment is decontamination, but, ironically, cleaning of sewage can also be smelly. Biosolids, the sludge that results from the treatment, still smells unpleasant.

What exactly is an odor?

An odor is a sensation registered via the sense of smell. The material that gives rise to an odor is called an odorant. It is useful to make this distinction between the sensation and the physical cause of the sensation. An odorant is a chemical, characteristically an organic chemical, that can become airborne. Odorants can trigger pleasant sensations, such as that of a baking cake or of a scented candle, neutral sensations, such as that of wood or rubbing alcohol, and unpleasant sensations, such as that of rotting fish or skunk. When the airborne molecules of odorant reach the receptors for smell, they are treated without respect to pleasantness or unpleasantness. That dimension enters the picture later, in the brain.

Why are some odors bad smelling?

Bad odors generally signal the condition of their sources. They convey the message that we should be wary of contact with their sources. Such sources often involve the activity of microorganisms, the scavengers of organic matter. In their scavenging, the microorganisms convert bigger molecules into smaller ones that can become airborne and can stimulate the sense of smell. As it turns out, the particular molecules created by the activity of microorganisms have extraordinary ability to stimulate smell. That is, it takes truly miniscule quantities of such molecules to stimulate smell. A thimbleful would often provide enough material to stimulate the sense of smell of every person on the face of the earth.

Do the chemicals that trigger bad odors ever come from sources that are not bad?

Indeed they do. Rarely do the odors of items in the real world, including the flavors of foods, come from just a single odorant or even just a few. Commonly, they come from dozens or even hundreds of odorants mixed in a certain proportion. Both the ingredients themselves and their proportions determine the character we perceive, such as a French-roasted coffee aroma or "eau de compost."

The odorants created by the activity of microorganisms appear in many products. In some cases they come from the actual activity of microorganisms, such as in cheeses, and in other cases they come from another source, such as cooking. When scientists have sought to decode the aroma of products that many people desire, such as orange peel, chocolate, grilled meat, beer, and hundreds of others, they have found numerous instances of malodorants. These often play essential roles in the appeal of the aromas. Although the palette for the aromas of everyday life has many thousands of ingredients, many of the same show up in both the good smelling and the bad smelling. The bad smelling ingredients do not just lurk as hidden contaminants beneath the surface, they actually make a positive contribution. For instance, the rind of an orange would not have its particular character without the contribution of some fecal-smelling odorants.

Why does it seem so compelling that the messenger is bad, if just the source is the material to avoid?

The sense of smell performs in two modes. In the first, it tells us about what exists at a distance, such as a field of fresh-mown hay. In the second, it tells us about what we have in our mouths, such as a piece of meat. In the second mode, smell works along with taste (sweet, sour, salty, and bitter) to tell us what we are about to swallow. Taste alone will not tell us that the meat has spoiled. The olfactory component of flavor signals that. For items in the mouth, the source and the messenger exist so close together that the distinction becomes irrelevant. The protection against eating contaminated food is so important that this mode dominates the interpretation of odors, even odors at a distance.

Are bad-smelling odorants not bad to breathe?

As a system specialized to detect minute amounts of chemical, the sense of smell generally can register the presence of a chemical (odorant) at a lower concentration than will harm the body. So, very few odors are harmful to breathe at levels we can just smell. This rule does not, however, dispense scientists from studying the toxic properties of individual chemicals. When they have done so, they have discovered that the odor threshold may lie many, many times below the threshold for a toxic effect or not so many times below such a threshold. Because odor perception and toxicity operate by different rules, one cannot use odor as a guide to health effects, but must refer to the relevant scientific information.

The knowledge that bad-smelling odorants may appear in good-smelling sources reminds us that toxicity lies in the dose.

Do wastewater treatment facilities take the threshold for toxic effects into consideration?

State governments and the federal government devise rules to protect the population from chemical exposures that could cause various kinds of health effects (acute, chronic, carcinogenic). The rules cover chemical emissions from wastewater treatment facilities. Those facilities must monitor their own emissions and report instances of noncompliance, an issue taken very seriously by the agencies that regulate the wastewater treatment facilities. Experience shows that the emissions may fall within compliance regarding toxic effects but may still have odor.

Do odors cause illness?

The simple answer is that odors do not cause illness, but this statement requires explanation. Stimulation of the sense of smell can sometimes cause symptoms, but these are not signs of illness in healthy people. Only very strong odors, beyond the permissible by environmental standards, should ever cause symptoms, such as gagging. However, somewhat less strong odors may cause people with certain existing chronic conditions to feel worse. For example, some people with asthma may feel more difficulty breathing during periods of odor pollution. In such cases, anxiety from exposure to malodors may alter symptoms through hormonal means. So, healthy people will not become ill, but some ill people may feel worse.

APPENDIX B

DISPERSION MODELING

Odor Dispersion Modeling

Odor dispersion modeling was performed to compare the maximum odor concentration offsite in 50-meter increments up to 2000 meters away from a typical emission source. The most conservative dilution factors were determined, reflecting the worst case (stable, low dispersion) meteorological conditions. With low wind speed and relatively stable air, dilution effects were minimized, so that the maximum level of exposure to odor for communities living near biosolids compared to the exposure level of workers one meter away was determined.

The plot of dilution factors as a function of distance is shown as Figure B-1. The graph was developed using the U.S. EPA SCREEN 3 Model, Version date 96043. The simple terrain inputs for Figure B-1 were as follows:

Source Type	Area
Emission Rate (g/sec/m^2)	1.0000
Source Height (m)	0.0000
Length of Larger Side (m)	10.0000
Length of Smaller Side (m)	10.0000
Receptor Height (m)	0.0000
Urban/Rural Option	Rural

Concentrations of pollutants were calculated by the SCREEN 3 program at 1, 50, 100, 150, and 200 to 2000 meters (at 100-meter intervals) distance from the source. The dilution factors at each distance were calculated by dividing the concentration at one meter from the source by the corresponding concentration. The graph shows the most conservative, i.e., the lowest dilution factors. Thus, at 2000 meters distance out from the emission source the dilution factor is 1580, meaning that the concentration of the odorant compounds is 1/1580th of the concentration at a distance of 1 meter from the source.

Figure B-1. Odor Dilution vs. Distance from Source

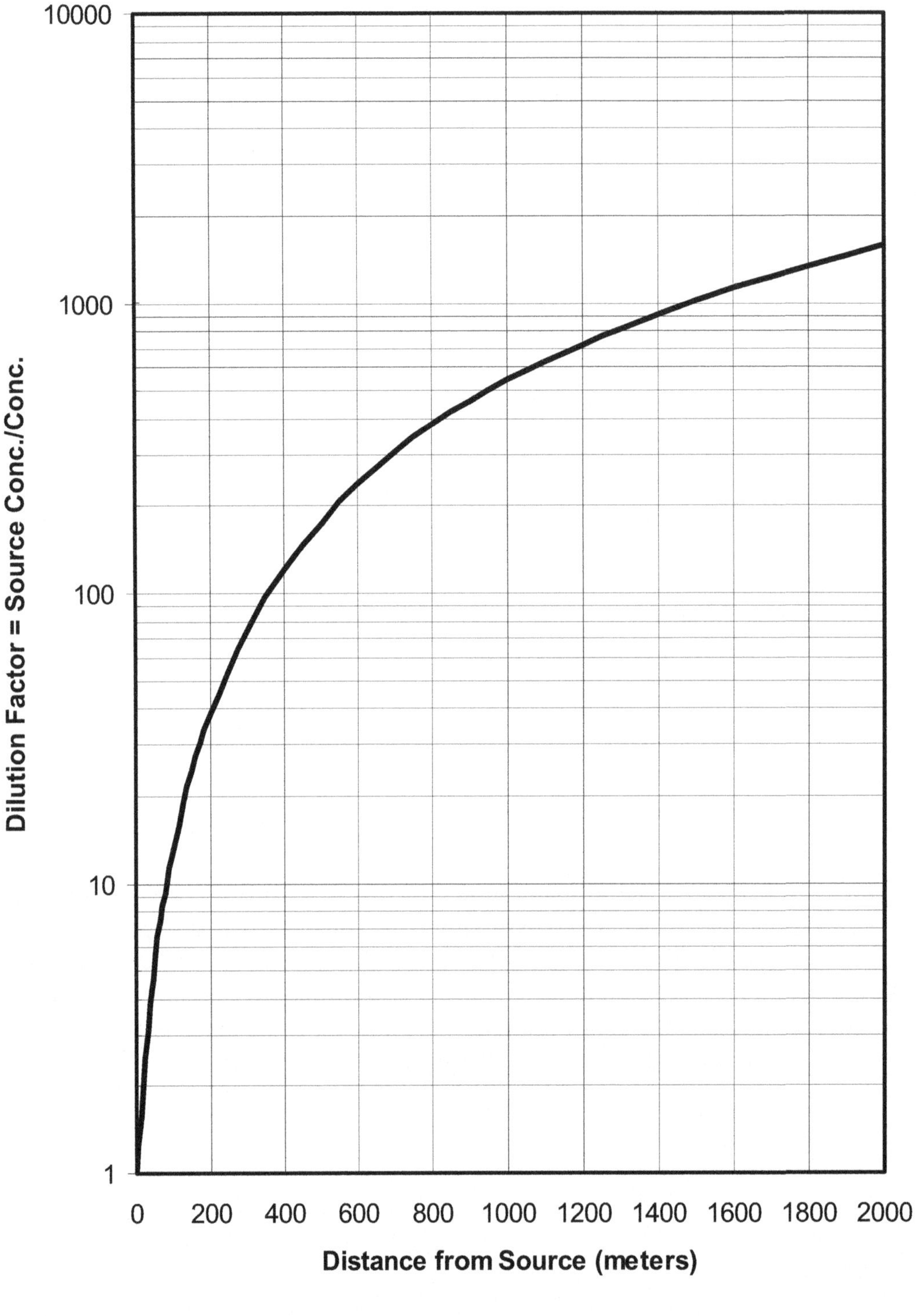

Note: Dilution factor equals one (1.0) at one (1.0) meter distance from source.

GLOSSARY OF TERMS

Amines	Class of organic compounds containing the $–NH_2$ group
Axon	The part of a nerve cell that conducts electrical activity toward another nerve cell
Chemesthetic threshold	Pertaining to the perception of feel from chemicals. Not far above the for feel, chemesthetic sensations may feel irritating or painful
Chemosensory	Pertaining to the perception of chemicals by smell, taste, or chemesthesis
Cilia	Finger-like projections from cells
Esters	Class of organic compounds containing two organic radicals joined by a –COO– group, usually the product of reaction between an organic acid and an organic alcohol
Glossopharyngeal	Pertaining to the ninth cranial nerve, which carries some information about taste and chemesthesis in the back of the oral cavity
Innervation	The property of having nerve supply
Ketones	Class of organic compounds containing two organic radicals joined by a –CO– group
Mercaptans	Class of organic compounds containing the -SH group
Mitral	Shaped like the tall pointed hat (miter) of a bishop and descriptive of the certain nerve cells in the chain of cells of the olfactory system
Mucosal	Pertaining to tissue with a layer of mucus, such as the tissue in the upper and lower airways
Nasopharynx	The structure that connects the back of the nasal cavity to the mouth
Neural	Pertaining to nerves or the nervous system
Neuron	A nerve cell
Olfaction	The sense of smell
Olfactory	Pertaining to the sense of smell
Organic acid	Class of organic compounds containing the –COOH (carboxyl) group, also carboxylic acids

Organic sulfides	Class of organic compounds containing the –S– (sulfur) element
Psychogenic	Originating from psychological causes
Pungency	Chemesthetic sensations arising from stimulation of the trigeminal nerve, particularly in the nose (nasal pungency)
Pyriform cortex	A receiving station for olfactory neural information
Quantitative structure-activity relationships (QSAR)	Relationships between a biological phenomenon, and the chemical or physical properties of the determinant of the phenomenon. A quantitative structure-activity relationship is normally expressed as an equation.
Somatogenic	Originating from the body, typically in contrast to originating from the mind
Somesthetic	Pertaining to the perception of feel, such as pressure, temperature, vibration, tickle, and pain
Transmembrane	Coursing through the wall (membrane) of a cell, as do certain protein molecules
Trigeminal	Pertaining to the fifth cranial nerve, which provides perception of feel to the face, eyes, and mucosal tissue in the upper airways and anterior oral cavity
Vagus	Pertaining to the tenth cranial nerve, which provides, among other things, perception of feel in the posterior oral cavity and throat

References

*Abraham, M.H., J.M.R. Gola, J.E., Cometto-Muñiz, and W.S. Cain. 2001. The correlation and prediction of VOC thresholds for nasal pungency, eye irritation and odour in humans. Indoor Built Environ 10 (3-4):252-257.

*Abraham, M.H., M. Hassanisadi, M. Jalali-Heravi, T. Ghafourian, W. S. Cain, and J. E. Cometto-Muñiz. 2003. Draize rabbit eye test compatibility with eye irritation thresholds in humans: a quantitative structure-activity relationship analysis. Toxicol Sci 76(2): 384- 391.

*Abraham, M.H., R. Kumarsingh, J.E. Cometto-Muñiz, and W.S. Cain. 1998a. An algorithm for nasal pungency thresholds in man. Arch Toxicol 72 (4):227-232.

*Abraham, M.H., R. Kumarsingh, J.E. Cometto-Muñiz, and W.S. Cain. 1998b. Draize eye scores and eye irritation thresholds in man combined into one quantitative structure-activity relationship. Toxicol in Vitro 12 (4):403-408.

Alix, C.M. 1998. Retrofits curb biosolids composting odors. BioCycle 39 (6):37-39.

Amoore, J.E. and E. Hautala. 1983. Odor as an aid to chemical safety: odor thresholds compared with threshold limit values and volatilities for 214 industrial chemicals in air and water dilution. J Appl Toxicol 3:272-290.

Anonymous. 1980. Reports of studies on the measurement of offensive odors (from 1972-1980) (Jap.). Tokyo: Japanese Environmental Agency.

Anonymous. 1999. Reducing odor and VOC emissions. BioCycle, March 1999, 68-72.

Assmuth, T. and K. Kalevi. 1992. Concentrations and toxicological significance of trace organic compounds in municipal solid waste landfill gas. Chemosphere 24 (9):1207-1216.

Attwood, P., P. Versloot, D. Heederik, R. De Witt, and J.S.M. Boleij. 1986. Assessment of dust and endotoxin levels in the working environment of Dutch pig farmers: a preliminary study. Am Occup Hyg 30 (2):201-208.

Bazemore, R., C.J. Wysocki, S. Murray, H.J. Lawley, and G. Preti. 2000. Amelioration of odorous components in spent mushroom compost. J Agric Food Chem 48:3694-3697.

Bazemore, R., C.J. Wysocki, P. Pitcher, T. Parsons, H.J. Lawley, L. Connolly, J. Louie, S. Murry, and G. Preti. 2000. Ameliorating swine slurry odors: an analytical and sensory approach. Chem Senses 25 (5):611 (abstract).

Beauchamp, R.O., Jr., J.S. Bus, J.A. Popp, C.J. Boreiko, and D.A. Andjelkovich. 1984. A critical review of the literature on hydrogen sulfide toxicity. CRC Crit Rev Toxic 13 (1):25-97.

Bhambhani, Y., R. Burnam, G. Snydmiller, I. MacLean, and R. Lovlin. 1996. Effects of 10-ppm hydrogen sulfide inhalation on pulmonary function in healthy men and women. J Occup Environ Med 38 (10):1012-1017.

Bhambhani, Y., R. Burnam, G. Snydmiller, I. MacLean, and T. Martin. 1994. Comparative physiological responses of exercising men and women to 5 ppm hydrogen sulphide exposure. Am Ind Hyg Assoc J 55 (11):1030-1035.

*Bonnin, C., A. Laborie, and H. Paillard. 1990. Odor nuisances created by sludge treatment: problems and solutions. Water Sci Tech 22 (12):65-74.

*Bottcher, R.W. 2001. An environmental nuisance: odor concentrated and transported by dust. Chem Senses 26 (3):327-31.

Brennan, B. 1993. Odour nuisance. Water Waste Treat 36:30-33.

*Cain, W.S. 1988. Olfaction. Stevens' Handbook of Experimental Psychology, Vol. 1: Perception and Motivation, rev. ed., edited by R.C. Atkinson, R.J. Herrnstein, G. Lindzey, and R.D. Luce. New York: Wiley.

*Cain, W.S., and F. Johnson, Jr. 1978. Lability of odor pleasantness: Influence of mere exposure. Perception 7:459-465.

*California Air Pollution Control Officers Association (CAPCOA). Toxics Committee. 1993. CAPCOA Air Toxics "Hot Spots" Program. State of California—Environmental Protection Agency.

Carpenter, A. and N. Beecher. 1997. Wood ash finds niche in biosolids composting. BioCycle, January 1997, 37-39.

*Cederlöf, R., L. Friberg, E. Jonsson, L. Kaij, and T. Lindvall. 1964. Studies of annoyance connected with offensive smell from a cellulose factory. Nord Hyg Tids 45:39-48.

Chang, C.W., H. Chung, C.F. Huang, and H.J.J. Su. 2001. Exposure assessment to airborne endotoxin, dust, ammonia, hydrogen sulfide, and carbon dioxide in open style swine houses. Annu Occup Hyg 45 (6):457-465.

Chen, Y., D.S. Bundy, and S.J. Hoff. 1999. Using olfactometry to measure intensity and threshold dilution ratio for evaluating swine odor. J Air Waste Manag Assoc 49 (7):847-53.

*Cometto-Muñiz, J.E. 2001. Physicochemical basis for odor and irritation potency of VOCs. In Indoor Air Quality Handbook, edited by J.D. Spengler, J. Samet and J.F. McCarthy. New York: McGraw-Hill.

*Cometto-Muñiz, J.E., and W.S. Cain. 1994. Perception of odor and nasal pungency from homologous series of volatile organic compounds. Indoor Air 4 (3):140-145.

*Cometto-Muñiz, J.E., and W.S. Cain. 1995. Relative sensitivity of the ocular trigeminal, nasal trigeminal and olfactory Systems to airborne chemicals. Chem Senses 20 (2):191-198.

*Cometto-Muñiz, J.E, and W.S. Cain. 1996. Physicochemical determinants and functional properties of the senses of irritation and smell. In Indoor Air and Human Health, 2nd Edition, edited by R.B. Gammage and B.A. Berven. Boca Raton: CRC Lewis Publishers.

*Cometto-Muñiz, J.E., W.S. Cain, M.H. Abraham, and J.M.R. Gola. 1999. Chemosensory detectability of 1-butanol and 2-heptanone singly and in binary mixtures. Physiol Behav 67 (2):269-276.

*Cometto-Muñiz, J.E., W.S. Cain, M.H. Abraham, and J.M.R. Gola. 2001. Ocular and nasal trigeminal detection of butyl acetate and toluene presented singly and in mixtures. Toxicol Sci 63 (2):233-244.

*Cometto-Muñiz, J.E., W.S. Cain, M.H. Abraham, and R. Kumarsingh. 1998. Trigeminal and olfactory chemosensory impact of selected terpenes. Pharmacol Biochem Behav 60 (3):765-770.

*Cometto-Muñiz, J.E., W.S. Cain, and H.K. Hudnell. 1997. Agonistic sensory effects of airborne chemicals in mixtures: Odor, nasal pungency, and eye irritation. Percept Psychophys 59 (5):665-674.

*Deane, M., and G. Sanders. 1977. Health effects of exposure to community odors from pulp mills, Eureka, California 1971. Environ Res 14 (1):164-81.

*Deane, M., and G. Sanders. 1978. Annoyance and health reactions to odor from refineries and other industries in Carson, California 1972. Environ Res 15 (1):119-32.

*Deane, M., G. Sanders, and E. Jonsson. 1977. Trends and community annoyance reactions to odors from pulp mills Eureka, California 1969-1971. Environ Res 14 (2):232-44.

*Defoer, N., I. De Bo, H. Van Langenhove, J. Dewulf, and T. Van Elst. 2002. Gas chromatography-mass spectrometry as a tool for estimating odour concentrations of biofilter effluents at aerobic composting and rendering plants. J Chromatogr A 970 (1-2):259-73.

Devai, I., and R.D. DeLaune. 2000. Emissions of reduced gaseous sulfur compounds from wastewater sludge: redox effects. Environ Eng Sci 17 (1):1-8.

*Devai, I., and R.D. DeLaune. 1999. Emission of reduced malodorous sulfur gases from wastewater treatment plants. Water Environ Res 71:203-208.

*Devos, M., J. Patte, J. Rouault, P. Laffort, and L.J. Van Gemert. 1990. Standardized Human Olfactory Thresholds. Oxford: Oxford University Press.

Donham, K.J., P. Haglind, Y. Peterson, R. Rylander, and L. Belin. 1989. Environmental and health studies of farm workers in Swedish swine confinement buildings. Brit J Ind Med 46:31-37.

Donham, K.J., and W.J. Popendorf. 1985. Ambient levels of selected gases inside swine confinement buildings. Am Ind Hyg Assoc J 46 (11):658-661.

Donham, K.J., M. Rubino, T.D. Thedell, and J. Kammermeyer. 1977. Potential health hazards to agricultural workers in swine confinement buildings. J Occup Med 19 (6):383-387.

Einarsen, A.M., A. Æsøy, A.I. Rasmussen, S. Bungum, and M. Sveberg. 2000. Biological prevention and removal of hydrogen sulphide in sludge at Lillehammer wastewater treatment plant. Water Sci Tech 41:175-182.

*Eriksson, N.E., O. Lowhagen, J.E. Nilsson, K. Norrlind, and J.A. Wihl. 1987. Flowers and other trigger factors in asthma and rhinitis — an inquiry study. Allergy 42:374-381.

Ettala, M., P. Rahkonen, V. Kitunen, O. Valo, and M. Salkinoja-Salonen. 1988. Quality of refuse, gas and water at a sanitary landfill. Aqua Fennica 18 (1):15-28.

*Finger, T.E., W.D. Silver, and D. Restrepo. 2000. Neurobiology of Taste and Smell. New York: Wiley-Liss.

*Flesh, R.D., J.C. Burns, and A. Turk. 1974. An evaluation of community problems caused by industrial odors. In Human Responses to Environmental Odors, edited by A. Turk, J.W. Johnston, Jr., and D.G. Moulton. New York: Academic Press.

*Floriano, W.B., N. Vaidehi, W.A. Goddard III, M.S. Singer, and G.M. Shepherd. 2000. Molecular mechanisms underlying differential odor responses of a mouse olfactory receptor, Proc Nat Acad Sci 97:10712-10716.

Georgieff, D., and T. Turnovska. 1999. Influence of odour emissions from cellulose-paper industry on some health indicators. Folia Med (Plovdiv) 41 (1):38-42.

Gostelow, P., and S.A. Parsons. 2000. Sewage treatment works odour measurement. Water Sci Tech 41 (6):33-40.

*Gostelow, P., S.A. Parsons, and R.M. Stuetz. 2001. Odor measurements for sewage treatment works. Water Res 35 (3):579-597.

*Grosser, K., R. Oelkers, T. Hummel, G. Geisslinger, K. Brune, G. Kobal and J. Lötsch. 2000. Olfactory and trigeminal event-related potentials in migraine. Cephalalgia, 20:621-631.

Guidotti, T.L. 1996. Hydrogen sulphide. J Occup Med 46 (5):367-371.

Haglind, P., and R. Rylander. 1987. Occupational exposure and lung function measurements among workers in swine confinement buildings. J Occup Med 29 (11):904-907.

Hall, G., and J. Andersson. 1983. Volatile fat oxidation products. I. determination of odour thresholds and odour intensity functions by dynamic olfactometry. Lebensm.-Wiss. u.-Technol 16:354-361.

Hammond, E.G., C. Fedler, and R.J. Smith. 1981. Analysis of particle-borne swine house odors. Agric Environ 6:395-401.

Harkov, R., S.J. Gianti, Jr., J.W. Bozzelli, and J. E. LaRegina. 1985. Monitoring volatile organic compounds at hazardous and sanitary landfills in New Jersey. J Environ Sci Health A20 (5):491-501.

Hartung, J., and V.R. Phillips. 1994. Control of gaseous emissions from livestock buildings and manure stores. J Agric Eng Res 57:173-189.

Heederik, D., R. Brouwer, K. Biersteker, and J.S.M. Boleij. 1991. Relationship of airborne endotoxin and bacteria levels in pig farms with the lung function and respiratory symptoms of farmers. Int Arch Occup Environ Health 62:595-601.

*Heida, H., F. Bartman, and S.C. Van der Zee. 1995. Occupational exposure and indoor air quality monitoring in a composting facility. Am Ind Hyg Assoc J 56 (1):39-43.

Hentz, L.H., Jr. 1997. The chemical, biological, and physical origins of biosolids emissions: a review. Paper read at Water Environment Federation Odors and VOC Emissions Speciality Conference.

Hentz, L.H., Jr. and Al Cassel. 2000. The effects of liquid sludge storage on biosolids odor emissions. Paper read at Water Environment Federation 14th Annual Residuals and Biosolids Management Conference.

Hentz, L.H., Jr., C.M. Murray, J.L. Thompson, L.L. Gasner, and J.B. Dunson, Jr. 1992. Odor control research at the Montgomery County regional composting facility. Water Environ Res 64 (1):13-18.

Hentz, L.H., Jr., W.E. Toffey, and C.E. Schmidt. 1996. Understanding the synergy between composting and air emissions. BioCycle, March 1996, 67-75.

Herr, C., A. zur Nieden, M. Mann, R.H. Boedeker, U. Gieler, and Th. Eikmann. 2001. Health effects of environmental odours and bioaerosols due to composting. Chem Senses 26 (6):736 (abstract).

Hobbs, P.J., T.H. Misselbrook, and B.F. Pain. 1995. Assessment of odours from livestock wastes by a photoionization detector, an electronic nose, olfactometry and gas chromatography-mass spectrometry. J Agric Eng Res 60:137-144.

Hobbs, P.J., T.H. Misselbrook, and B.F. Pain. 1997. Characterization of odorous compounds and emissions from slurries produced from weaner pigs fed dry feed and liquid diets. J Sci Food Agric 73:437-445.

Hobbs, P.J., B.F. Pain, R.M. Kay, and P.A. Lee. 1996. Reduction of odorous compounds in fresh pig slurry by dietary control of crude protein. J Sci Food Agric 71:508-514.

*Horst, W.G., F. Mattern, S.H. Vold, and J.M. Walker. 1991. Controlling compost odors. Part II. BioCycle 32:46-51.

Islam, A.K.M.N., K. Hanaki, and T. Matsuo. 1998. Fate of dissolved odorous compounds in sewage treatment plants. Water Sci Tech 38 (3):337-344.

Jaakkola, J.J.K., V. Vilkka, O. Marttila, P. Jäppinen, and T. Haahtela. 1990. The effects of malodorous sulfur compounds from pulp mills on respiratory and other symptoms. Am Rev Respir Dis 142:1334-1350.

Jäppinen, P., J. Kangas, L. Silakoski, and H. Savolainen. 1993. Volatile metabolites in occupational exposure to organic sulfur compounds. Toxicology 67:104-106.

Jäppinen, P., and R. Tenhunen. 1990. Hydrogen sulphide poisoning: blood sulphide concentration and changes in haem metabolism. Brit J Ind Med 47:283-285.

*Jonsson, E. 1974. Annoyance reactions to environmental odors. In Human Responses to Environmental Odors, edited by A. Turk, J. Johnston, J.W. and D.G. Moulton. New York: Academic Press.

*Jonsson, E., M. Deane, and G. Sanders. 1975. Community reactions to odors from pulp mills: a pilot study in Eureka, California. Environ Res 10 (2):249-70.

Junker, M.H., B. Danuser, C. Monn, and T. Koller. 2001. Acute sensory responses of nonsmokers at very low environmental tobacco smoke concentrations in controlled laboratory settings. Environ Health Perspect 109 (10):1045-52.

Kangas, J., P. Jappinen, and H. Savolainen. 1984. Exposure to hydrogen sulfide, mercaptans, and sulfur dioxide in pulp industry. Am Ind Hyg Assoc J 45 (12):787-790.

*Kangas, J., A. Nevalainen, A. Manninen, and H. Savolainen. 1986. Ammonia, hydrogen sulphide and methyl mercaptides in Finnish municipal sewage plants and pumping stations. Sci Tot Environ 57:49-55.

*Katon, W., M. Sullivan, and E. Walker. 2001. Medical symptoms without identified pathology: Relationship to psychiatric disorders, childhood and adult trauma, and personality traits. Ann Intern Med 134:917-925.

Kaye, R., and K. Jiang. 2000. Development of odour impact criteria for sewage treatment plants using odour complaint history. Water Sci Tech 41 (6):57-64.

Khan, A.A., M.M. Schuler, M.G. Prior, S. Yong, R.W. Coppok, L.Z. Florence, and L.E. Lillie. 1990. Effects of hydrogen sulphide exposure on Jung mitochondrial respiratory chain enzymes in rats. Toxicol Appl Pharm 103:482-490.

Kiviranta, H., A. Tuomainen, M. Reiman, S. Laitinen, A. Nevalainen, and J. Liesivouri. 1999. Exposure to airborne microorganisms and volatile organic compounds in different types of waste handling. Annu Agric Environ Med 6:39-44.

Koe, L.C.C. 1985. Hydrogen sulphide odor in sewage atmospheres. Water Air Soil Pollut 24:297-306.

Krugel, S., L. Nemeth, and C. Peddie. 1998. Extending thermophilic anaerobic digestion for production biosolids at the Greater Vancouver Regional Districts Annacis Island Wastewater Treatment Plant. Water Sci Tech 38 (8-9):409-416.

Krzymien, M., M. Day, K. Shaw, and L. Zaremba. 1999. An Investigation of odors and volatile organic compounds released during composting. J Air Waste Manag Assoc 49 (7):804-813.

*Lambert, S.D., A.L. Beaman, and P. Winter. 2000. Olfactometric characterization of sludge odours. Water Sci Tech 41:49-55.

Lambert, S.D., and K.E. McGrath. 2000. Can stored sludge cake be deodorised by chemical or biological treatment? Water Sci Tech 41:133-139.

*Laube, B.L., B.A. Curbow, R.W. Costello, and S.T. Fitzgerald. 2002. A pilot study examining the relationship between stress and serum cortisol concentrations in women with asthma. Respir. Medicine 96:823-828.

Leonardos, G., D. Kendall, and N. Barnard. 1969. Odor threshold determinations of 53 odorant chemicals. J Air Pollut Control Assoc 19 (2):91-95.

Lewis, D.L., D.K. Gattie, M.E. Novak, S. Sanchez, and C. Pumphrey. 2002. Interactions of pathogens and irritant chemicals in land-applied sewage sludges (biosolids). BMC Public Health 2 (1):11.

*Lipscomb, J.A., L.R. Goldman, K.P. Satin, D.F. Smith, W.A. Vance, and R.R. Neutra. 1991. A follow-up study of the community near the McColl waste disposal site. Environ Health Perspect 94:15-24.

Logue, J.N., K. Ramaswamy, and J.H. Hersh. 2001. Investigation of illness associated with exposure to hydrogen sulfide among Pennsylvania school students. J Environ Health 63 (6):9-13.

Louhelainen, K., J. Kangas, A. Veijanen, and P. Viilos. 2001. Effect of in situ composting on reducing offensive odors and volatile organic compounds in swineries. Am Ind Hyg Assoc J 62 (2):159-167.

*Luginaah, I.N., S.M. Taylor, S.J. Elliott, and J.D. Eyles. 2000. A longitudinal study of the health impacts of a petroleum refinery. Soc Sci Med 50 (7-8):1155-66.

Lunn, F., and J. Van de Vyver. 1977. Sampling and analysis of air in pig houses. Agric Environ 3:159-169.

Mahin, T., R. Pope, and C. McGinley. 2000. When is a smell a nuisance? An overview of different approaches taken around the world in setting odor-control regulations. Water Environ Tech 12 (5):49-53.

*Mahin, T.D. 2001. Comparison of different approaches used to regulate odours around the world. Water Sci Technol 44 (9):87-102.

Mannebeck, D., and H. Mannebeck. 2001. Interlaboratory comparison of dynamic olfactometry in Central Europe 2000. Water Sci Technol 44 (9):27-32.

McIntyre, A. 2000. Application of dispersion modeling to odour assessment: a practical tool or a complex trap? Water Sci. Tech. 41 (6):81-88.

Michel, F.C., Jr., and C.A. Reddy. 1998. Effect of oxygenation level on yard trimmings composting rate, odor production, and compost quality in bench-scale reactors. Compost Sci Utiliz 6:6-14.

Miller, F.C., and B.J. Macauley. 1988. Odours arising from mushroom composting: a review. Aust J Exp Agric 28:553-560.

*Millqvist, E., and O. Lowhagen. 1996. Placebo-controlled challenges with perfume in patients with asthma-like symptoms. Allergy 51:434-439.

*Millqvist, E., and O. Lowhagen. 1998. Methacholine provocations do not reveal sensitivity to strong scents. Ann Allergy Asthma Immunol 80:381-384.

*Muirhead, T., P. LaFond, and D. Dennis. 1993. Air handling and scrubber retrofits optimize odor control. BioCycle 34:68-75.

*National Research Council (NRC). 2002. Biosolids Applied to Land: Advancing Standards and Practices. Washington: The National Academies Press.

Noble, R., P.J. Hobbs, A. Dobrovin-Pennington, T.H. Misselbrook, and A. Mead. 2001. Olfactory response to mushroom composting emissions as a function of chemical concentration. J Envrion Qual 30:760-767.

O'Neill, D.H., and V.R. Phillips. 1992. A review of the control of odour nuisance from livestock buildings: part 3, properties of the odorous substances which have been Identified in livestock wastes or in the air around them. J Agric Eng Res 53:23-50.

*Ottoson, D. 1983. Physiology of the Nervous System. New York: Oxford University Press.

Pain, B.F., C.R. Clarkson, V.R. Phillips, J.V. Klarenbeek, T.H. Misselbrook, and M. Bruins. 1991. Odour emission arising from application of livestock slurries on land: measurements following spreading using a micrometeorological technique and olfactometry. J Agric Eng Res 48:101-110.

Paulsrud, B., and K.T. Nedland. 1997. Strategy for land application of sewage sludge in Norway. Water Sci Tech 36:283-290.

Pawlowski, R. 1992. Case study of confined-space death illustrates need for written procedure. Occup Health Safety 61 (9):26-27.

Reiffenstein, R.J., W.C. Hulbert, and S.H. Roth. 1992. Toxicology of hydrogen sulphide. Annu Rev Pharm Toxicol 32:109-134.

Reynolds, S.J., K.J. Donham, P. Whitten, J.A. Merchant, L.F. Burmesiter, and W.J. Popendorf. 1996. Longitudinal evaluation of dose-response relationships for environmental exposures and pulmonary function in swine production workers. Am J Ind Med 29:33-40.

*Rosenfeld, P.E., C.L. Henry, and D. Bennet. 2001. Wastewater dewatering polymer affect on biosolids. Water Environ Res 73 (3):363-367.

*Ruth, J.H. 1986. Odor thresholds and irritation levels of several chemical substances: a review. Am Ind Hyg Assoc J 47:142-151.

Sarkar, U., and S.E. Hobbs. 2002. Odour from municipal solid waste (MSW) landfills: a study on the analysis of perception. Environ Int 27 (8):655-62.

Schiffman, S.S., J. Bennett, and J.H. Raymer. 2001. Measurement of odor and odorants from swine facilities. Chem Senses 26 (6):735 (abstract).

*Schiffman, S.S., J.M. Walker, P. Dalton, T.S. Lorig, J.H. Raymer, D. Shusterman, and C. Williams. 2000. Potential health effects of odor from animal operations, wastewater treatment, and recycling of byproducts. J Agromed 7 (1):7-81.

Schiffman, S.S., J.M. Walker, R. Small, and P.D. Millner. 1998. Workshop on health effects of odors. Paper read at Animal Production Systems and the Environment: An International Conference on Odor, Water Quality, Nutrient Management and Socioeconomic Issues, at Des Moines, Iowa State University.

Schmidt, C.E., E. Winegar, and L.H. Hentz, Jr. 1997. Assessment of odorous air emissions from compost facilities. Paper read at Air and Waste Management Association's 90th Annual Meeting and Exhibition, June 8-13, 1997, at Toronto, Ontario, Canada.

*Segala, C., D. Poizeau, and J.M. Mace. 2003. [Odors and health: a descriptive epidemiological study around a wastewater treatment plant.] Rev Epidemiol Santé Publique 51:210-214.

*Shim, C., and M.H. Williams, Jr. 1986. Effect of odors in asthma. Am J Med 80:18-22.

*Shusterman, D. 1992. Critical Review: The health significance of environmental odor pollution. Arch Environ Health 47:76-87.

*Shusterman, D. 1999. The health significance of environmental odour pollution: Revisited. J Environ Med 1 249-258.

*Shusterman, D., J. Lipscomb, R. Neutra, and K. Satin. 1991. Symptom prevalence and odor-worry interaction near hazardous waste sites. Environ Health Perspect 94:25-30.

*Smailing, K.B., P.E. McKnight, and N. Afari. 2002. A prospective study of mood and stress to pulmonary function among patients with asthma. J Asthma 39:510-510.

Snyder, J.W., E.F. Safir, G.P. Summerville, and R.A. Middleberg. 1995. Occupational fatality and persistent neurological sequelae after mass exposure to hydrogen sulfide. Am J Emerg Med 13 (2):199-203.

*Søstrand, P., B. Tvedt, W. Eduard, E. Bye, and K. Heldal. 2000. Hazardous peak concentrations of hydrogen sulfide gas related to the sewage purification process. Am Ind Hyg Assoc J 61 (1):107-110.

*Steinheider, B. 1998/9. Environmental odours and somatic complaints. Zbl Hyg Umweltmed 202 (2-4):101-119.

*Steinheider, B., R. Both, and Winnecke, G. 1998a. Field studies on environmental odors inducing annoyance as well as gastric and general health-related symptoms. J Psychophysiol, Supplement 64-79.

*Steinheider, B., R. Both, and G. Winnecke. 1998b. Die Erfassung der Geruchsbelästigung durch Tierstallimmissionen bei Anworhern. (Assessment of odor annoyance in the vicinity of livestock buildings: An effect-related comparison of exposure assessment by either dispersion or systematic field observations.). Gefahrstoffe - Reinhaltung der Luft 58:411-416.

Sutton, A.L., K.B. Kephart, M.W.A. Verstegen, T.T. Canh, and P.J. Hobbs. 1999. Potential for reduction of odorous compounds in swine manure through diet modification. J Anim Sci 77:430-439.

*Terkel, S. 1970. Working: People Talk About What They Do All Day and How They Feel About What They Do. New York: The New Press.

Thu, K. 1998. Odor problems from large-scale agriculture: nuisance or public health problem. Health Environ Digest 12 (8):57-59.

Thu, K., R. Donham, R. Ziegenhorn, S. Reynolds, P.S. Thorne, P. Subramanian, P. Whitten, and J. Stookesberry. 1997. A control study of the physical and mental health of residents living near a large-scale swine operation. J Agric Safety Health 3 (1):13-26.

Tkach. 1965. Combined effect of acetone and acetophone in the atmosphere. Hyg Sanit USSR 30:179.

*Tolvanen, O.K., K.I. Hänninen, A. Veijanen, and K. Villberg. 1998. Occupational hygiene in biowaste composting. Waste Manag Res 16:525-540.

Van Durme, G.P., B.F. McNamara, and C.M. McGinley. 1992. Benchscale removal of odor and volatile organic compounds at a composting facility. Water Environ Res 64:19-27.

Van Gemert, L.J. 1984. Compilation of odour threshold values in air, Division of Nutrition and Food Research TNO. The Netherlands: Zeist.

Van Gemert, L.J., and A.H. Nettenbreijer. 1977. Compilation of odour threshold values in air and water. The Netherlands: Zeist.

Van Langenhove, H., K. Roelstraete, N. Shamp, and J. Houtmeyers. 1985. GC-MS identification of odorous volatiles in wastewater. Water Res 19 (5):597-603.

Vandergheynst, J.S., D.J. Cogan, P.J. Defelice, J.M. Gossett, and L.P. Walker. 1998. Effect of process management on the emissions of organosulfur compounds and gaseous antecedents from composting processes. Environ Sci Technol 32 (23):3713-3718.

*Wacogne, C., J.P. Lacoste, E. Guillibert, F.C. Hugues and C. Le Jeunne. 2003. Stress, anxiety, depression and migraine. Cephalalgia 23:451-455.

Wakefield, S., and S.J. Elliott. 2000. Environmental risk perception and well-being: effects of the landfill siting process in two southern Ontario communities. Soc Sci Med 50 (7-8):1139-54.

Wakefield, S., and S.J. Elliott. 2000. Environmental risk perception and well-being: effects of the landfill siting process in two southern Ontario communities. Soc Sci Med 50:1139-1154.

Watt, M.M., S.J. Watt, and A. Seaton. 1997. Episode of toxic gas exposure in sewer workers. Occup Environ Med 54:277-280.

Whorton, J. 2001. "The insidious foe"—sewer gas. West J Med 175 (6):427-8.

Wilby, F.V. 1969. Variation in recognition odor threshold of a panel. J Air Pollut Control Assoc 19 (2):96-100.

Williams, T.O. 1995. Odors and VOC emissions control methods. BioCycle 36:49-56.

Winnecke, G., R. Both, and K. Sucker. 2001. Adverse effects of environmental odors: annoyance responses and symptom reporting. Chem Senses 26 (6):734 (abstract).

*Winnecke, G., and J. Kastka. 1977. Odour pollution and odour annoyance reactions in industrial areas of the Rhine-Ruhr region. In: Proceedings of the Sixth International Symposium on Olfaction and Taste, edited by J. L. Magnen and P. M. Leod. London: Information Retrieval Limited.

*Winnecke, G., and J. Kastka. 1987. Comparison of odour-annoyance data from different industrial sources: Problems and implications. In Environmental Annoyance: Characterization, Measurement, and Control, edited by H. S. Koelega. Amsterdam: Elsevier Science Publishers.

*Winnecke, G., M. Neuf, and B. Steinheider. 1996. Separating the impact of exposure and personality in annoyance response to environmental stressors, particularly odors. Environ Internat 22:73-81.

Winter, P., and P.S. Duckham. 2000. Analysis of volatile odour compounds in digested sewage sludge and aged sewage sludge cake. Water Sci Tech 41 (6):73-80.

Wright, L.L. 2001. Searching fee and non-fee toxicology information resources: an overview of selected databases. Toxicol 157:89-110.

Wysocki, C.J., G. Preti, R. Bazemore, T. Parsons, P. Pitcher, L. Connolly, and J. Louie. 2001. Agriculturally derived odors: irritants, cognitoxins or just plain nasty smells. Chem Senses 26 (6):735 (abstract).

Wysocki, C.J., G. Preti, R. Bazemore, P. Pitcher, T. Parsons, L. Connolly, and J. Louie. 2000. Feeding of swine to ameliorate odors: an analytical and sensory approach. Chem Senses 25 (5):611 (abstract).

Yasuhara, A., K. Fuwa, and M. Jimbu. 1984. Identification of odorous compounds in fresh and rotten swine manure. Agric Biol Chem 48 (12):3001-3010.

Young, P.J. 1984. Odours from effluent and waste treatment. Efflu Water Treat J 24:189-195.

Zahn, J.A., A.A. DiSpirito, Y.S. Do, B.E. Brooks, E.E. Cooper, and J.L. Hatfield. 2001. Correlation of human olfactory responses to airborne concentrations of malodorous volatile organic compounds emitted from swine effluent. J Environ Qual 30:624-634.

Zahn, J.A., J.L. Hatfield, Y.S. Do, A.A. DiSpirito, D.A. Laird, and R.L. Pfeiffer. 1997. Characterization of volatile organic emissions and wastes from a swine production facility. J Environ Qual 26:1687-1696.

Zahn, J.A., J.L. Hatfield, D.A. Laird, T.T. Hart, Y.S. Do, and A.A. DiSpirito. 2001. Functional classification of swine manure management systems based on effluent and gas emission characteristics. J Environ Qual 30:635-647.

Zejda, J.E., E. Barber, J.A. Dosman, S.A. Olenchok, H.H. McDuffie, C. Rhodes, and T. Hurst. 1994. Respiratory health status in swine producers relates to endotoxin exposure in the presence of low dust levels. J Occup Med 36 (1):49-56.

* Asterisked references were cited in the text. The other references were reviewed but not cited.

WASTEWATER UTILITY

Alabama
Montgomery Water Works & Sanitary Sewer Board

Alaska
Anchorage Water & Wastewater Utility

Arizona
Gila Resources

Glendale, City of, Utilities Department

Mesa, City of

Peoria, City of

Pima County Wastewater Management

Arkansas
Little Rock Wastewater Utility

California
Calaveras County Water District

Central Contra Costa Sanitary District

Contra Costa Water District

Crestline Sanitation District

Delta Diablo Sanitation District

Dublin San Ramon Services District

East Bay Dischargers Authority

East Bay Municipal Utility District

El Dorado Irrigation District

Fairfield-Suisun Sewer District

Fresno Department of Public Utilities

Irvine Ranch Water District

Las Virgenes Municipal Water District

Livermore, City of

Lodi, City of

Los Angeles, City of

Los Angeles County, Sanitation Districts of

Napa Sanitation District

Orange County Sanitation District

Palo Alto, City of

Riverside, City of

Sacramento Regional County Sanitation District

San Diego Metropolitan Wastewater Department, City of

San Francisco, City & County of

San Jose, City of

Santa Barbara, City of

Santa Rosa, City of

South Bayside System Authority

South Coast Water District

South Orange County Wastewater Authority

Sunnyvale, City of

Union Sanitary District

West Valley Sanitation District

Colorado
Boulder, City of

Colorado Springs Utilities

Littleton/Englewood Water Pollution Control Plant

Metro Wastewater Reclamation District, Denver

Connecticut
The Mattabasset District

New Haven, City of, WPCA

District of Columbia
District of Columbia Water & Sewer Authority

Florida
Broward, County of

Fort Lauderdale, City of

Gainesville Regional Utilities

JEA

Miami-Dade Water & Sewer Authority

Orange County Utilities Department

Orlando, City of

Reedy Creek Improvement District

Seminole County Environmental Services

St. Petersburg, City of

Stuart Public Utilities

Tallahassee, City of

Tampa, City of

Toho Water Authority

West Palm Beach, City of

Georgia
Atlanta Department of Watershed Management

Augusta, City of

Clayton County Water Authority

Cobb County Water System

Columbus Water Works

Fulton County

Gwinnett County Department of Public Utilities

Savannah, City of

Hawaii
Honolulu, City and County of

Illinois
American Bottoms Wastewater Treatment Plant

Greater Peoria Sanitary District

Kankakee River Metropolitan Agency

Metropolitan Water Reclamation District of Greater Chicago

North Shore Sanitary District

Wheaton Sanitary District

Iowa
Ames, City of

Cedar Rapids Wastewater Facility

Des Moines Metro Wastewater Reclamation Authority

Iowa City, City of

Kansas
Johnson County Unified Wastewater Districts

Unified Government of Wyandotte County/Kansas City, City of

Kentucky
Louisville & Jefferson County Metropolitan Sewer District

Louisiana
Sewerage & Water Board of New Orleans

Maine
Bangor, City of

Portland Water District

Maryland
Anne Arundel County Bureau of Utility Operations

Howard County Department of Public Works

Washington Suburban Sanitary Commission

Massachusetts
Boston Water & Sewer Commission

Upper Blackstone Water Pollution Abatement District

Michigan
Ann Arbor, City of

Detroit, City of

Holland Board of Public Works

Lansing, City of

Owosso Mid-Shiawassee County WWTP

Saginaw, City of

Wayne County Department of Environment

Wyoming, City of

Minnesota
Rochester, City of

Western Lake Superior Sanitary District

Missouri
Independence, City of

Kansas City Missouri Water Services Department

Little Blue Valley Sewer District

Nebraska
Lincoln Wastewater System

Nevada
Henderson, City of

New Jersey
Bergen County Utilities Authority

Passaic Valley Sewerage Commissioners

New York
New York City Department of Environmental Protection

Rockland County Solid Waste Management Authority/ Sewer District

North Carolina
Charlotte/Mecklenburg Utilities

Durham, City of

Metropolitan Sewerage District of Buncombe County

Orange Water & Sewer Authority

Ohio
Akron, City of

Butler County Department of Environmental Services

Columbus, City of

Metropolitan Sewer District of Greater Cincinnati

Northeast Ohio Regional Sewer District

Summit, County of

Oklahoma
Tulsa, City of

Oregon
Clean Water Services

Eugene/Springfield Water Pollution Control

Water Environment Services

Pennsylvania
Philadelphia, City of

University Area Joint Authority, State College

South Carolina
Charleston Commissioners of Public Works

Mount Pleasant Waterworks & Sewer Commission

Spartanburg Sanitary Sewer District

Tennessee
Cleveland, City of

Knoxville Utilities Board

Murfreesboro Water & Sewer Department

Nashville Metro Water Services

Texas
Austin, City of

Dallas Water Utilities

Denton, City of

El Paso Water Utilities

Fort Worth, City of

Gulf Coast Waste Disposal Authority

Houston, City of

San Antonio Water System

Trinity River Authority

Utah
Salt Lake City Corporation

Virginia
Alexandria Sanitation Authority
Arlington, County of
Fairfax County Virginia
Hampton Roads Sanitation District
Henrico, County of
Hopewell Regional Wastewater Treatment Facility
Loudoun County Sanitation Authority
Lynchburg Regional WWTP
Prince William County Service Authority
Richmond, City of
Rivanna Water & Sewer Authority

Washington
Edmonds, City of
Everett, City of
King County Department of Natural Resources
Seattle Public Utilities
Sunnyside, Port of
Yakima, City of

Wisconsin
Green Bay Metro Sewerage District
Madison Metropolitan Sewerage District
Milwaukee Metropolitan Sewerage District
Racine, City of
Sheboygan Regional Wastewater Treatment
Wausau Water Works

Australia
South Australian Water Corp.
Sydney Water Corp.
Water Corp. of Western Australia

Canada
Greater Vancouver Regional District
Toronto, City of, Ontario
Winnipeg, City of, Manitoba

Mexico
Servicios de Agua y Drenaje de Monterrey, I.P.D.

New Zealand
Watercare Services Limited

Singapore
Singapore Public Utilities Board

United Kingdom
Yorkshire Water Services Limited

STORMWATER UTILITY

California
Los Angeles, City of, Department of Public Works
Monterey, City of
San Francisco, City & County of
Santa Rosa, City of
Sunnyvale, City of

Colorado
Boulder, City of

Georgia
Griffin, City of

Iowa
Cedar Rapids Wastewater Facility
Des Moines Metro Wastewater Reclamation Authority

Kansas
Overland Park, City of

Kentucky
Louisville & Jefferson County Metropolitan Sewer District

Maine
Portland Water District

Minnesota
Western Lake Superior Sanitary District

North Carolina
Charlotte, City of, Stormwater Services

Pennsylvania
Philadelphia, City of

Tennessee
Chattanooga Storm Water Management

Washington
Seattle Public Utilities

STATE

Arkansas Department of Environmental Quality
Fresno Metropolitan Flood Control District, Calif.
Urban Drainage & Flood Control District, Colo.

CORPORATE

ADS Environmental Services
The ADVENT Group Inc.
Alan Plummer & Associates
Alden Research Laboratory
Aqua-Aerobic Systems, Inc.
Aquateam–Norwegian Water Technology Centre A/S
BaySaver Inc.
BioVir Laboratories, Inc.
Black & Veatch
Boyle Engineering Corporation
Brown & Caldwell
Burns & McDonnell
CABE Associates Inc.
The Cadmus Group
Camp Dresser & McKee Inc.
Carollo Engineers Inc.
Carpenter Environmental Associates Inc.
CDS Technologies Inc.
Chemtrac Systems Inc.
CH2M HILL
Damon S. Williams Associates, LLC
David L. Sheridan, P.C.
Dewling Associates, Inc.
Earth Tech Inc.
Eco-Matrix
Ecovation
Environmental Engineers International
EMA Inc.
The Eshelman Company Inc.
Finkbeiner, Pettis, & Strout Inc.
Freese & Nichols, Inc.
ftn Associates Inc.
Fuss & O'Neill, Inc.
Gannett Fleming Inc.
Golder Associates Inc.
Greeley and Hansen LLC
Hazen & Sawyer, P.C.
HDR Engineering Inc.
HNTB Corporation
HydroQual Inc.
Infilco Degremont Inc.
Ingersoll-Rand Energy Systems
Insituform Technologies Inc.
Institute for Environmental Technology & Industry, Korea
Jacobson Helgoth Consultants Inc.
Jason Consultants Inc.
Jordan, Jones, & Goulding Inc.
KCI Technologies Inc.
Kelly & Weaver, P.C.
Kennedy/Jenks Consultants
Komline Sanderson Engineering Corporation
Lawler, Matusky & Skelly Engineers, LLP
Limno-Tech Inc.
Lombardo Associates Inc.
Malcolm Pirnie Inc.
McKim & Creed
MEC Analytical Systems Inc.
Metcalf & Eddy Inc.
MWH
New England Organics
Odor & Corrosion Technology Consultants Inc. (OCTC)
Oswald Green, LLC
PA Government Services Inc.
Parametrix Inc.
Parsons
Post, Buckley, Schuh & Jernigan
R&D Engineering/Conestoga Rover & Associates
The RETEC Group
RMC, Inc. (Raines, Melton & Carella)
R.M. Towill Corporation
Ross & Associates Ltd.
Royce Technologies
SAIC Maritime Technical Group
Short Elliott Hendrickson, Inc.
Stantec Consulting Group, Inc.
Stormwater Management, Inc.
Synagro Technologies, Inc.
Tetra Tech Inc.
Trojan Technologies Inc.
URS Corporation
Veolia Water NATC
Wade-Trim Inc.
Weston Solutions Inc.
Woodard & Curran
WRc/D&B, LLC
WWETCO, LLC
Zoeller Pump Company

INDUSTRY

American Electric Power
ChevronTexaco Energy Research & Technology Company
The Coca-Cola Company
Dow Chemical Company
DuPont Company
Eastman Kodak Company
Eli Lilly & Company
Merck & Company Inc.
ONDEO Services
Procter & Gamble Company
PSEG Services Corp.
RWE Thames Water Plc
Severn Trent Services Inc.
United Water Services LLC

Note: List as of 7/1/04

WERF Product Order Form

As a benefit of joining the Water Environment Research Foundation, subscribers are entitled to receive one complimentary copy of all final reports and other products. Additional copies are available at cost (usually $10). To order your complimentary copy of a report, please write "free" in the unit price column. WERF keeps track of all orders. If the charge differs from what is shown here, we will call to confirm the total before processing.

Name _______________ Title _______________

Organization _______________

Address _______________

City _______________ State _______________ Zip Code _______________ Country _______________

Phone _______________ Fax _______________ Email _______________

Stock #	Product	Quantity	Unit Price	Total
			Postage & Handling	
			VA Residents Add 4.5% Sales Tax	
			Canadian Residents Add 7% GST	
			TOTAL	

Method of Payment:

(All orders must be prepaid.)

- ❑ Check or Money Order Enclosed
- ❑ Visa ❑ Mastercard ❑ American Express

Account No. _______________ Exp. Date _______________

Signature _______________

Shipping & Handling:

Amount of Order	United States	Canada & Mexico	All Others
Up to but not more than:	Add:	Add:	Add:
$20.00	$5.00*	$8.00	50% of amount
30.00	5.50	8.00	40% of amount
40.00	6.00	8.00	
50.00	6.50	14.00	
60.00	7.00	14.00	
80.00	8.00	14.00	
100.00	10.00	21.00	
150.00	12.50	28.00	
200.00	15.00	35.00	
More than $200.00	Add 20% of order	Add 20% of order	

*minimum amount for all orders

Note: Please make checks payable to the Water Environment Research Foundation.

To Order (Subscribers Only):

Log on to www.werf.org and click on the " Product Catalog."

Phone: (703) 684-2470
Fax: (703) 299-0742.

WERF
Attn: Subscriber Services
635 Slaters Lane
Alexandria, VA 22314-1177

To Order (Non-Subscribers):

Non-subscribers may be able to order WERF publications either through WEF (www.wef.org) or IWAP (www.iwapublishing.com). Visit WERF's website at www.werf.org for details.

IWA Publishing's authorised EU representative for General Product Safety Regulations is Diane D'Arras, 15 rue Duret, 75116 Paris, France, e-mail: safety@iwap.co.uk.

Printed and bound by CPI Group (UK) Ltd, Croydon, CR0 4YY
17/09/2026
02218813-0008